ALSO BY NATHALIA HOLT

*Wise Gals: The Spies Who Built the CIA
and Changed the Future of Espionage*

*The Queens of Animation: The Untold Story of the Women
Who Transformed the World of Disney and Made Cinematic History*

*Rise of the Rocket Girls: The Women Who Propelled Us,
from Missiles to the Moon to Mars*

Cured: The People Who Defeated HIV

THE BEAST IN THE CLOUDS

THE ROOSEVELT BROTHERS' DEADLY QUEST TO FIND THE MYTHICAL GIANT PANDA

NATHALIA HOLT

ONE SIGNAL
PUBLISHERS
ATRIA
New York   Amsterdam/Antwerp   London
Toronto   Sydney/Melbourne   New Delhi

ONE SIGNAL PUBLISHERS

ATRIA

An Imprint of Simon & Schuster, LLC
1230 Avenue of the Americas
New York, NY 10020

For more than 100 years, Simon & Schuster has championed authors and the stories they create. By respecting the copyright of an author's intellectual property, you enable Simon & Schuster and the author to continue publishing exceptional books for years to come. We thank you for supporting the author's copyright by purchasing an authorized edition of this book.

No amount of this book may be reproduced or stored in any format, nor may it be uploaded to any website, database, language-learning model, or other repository, retrieval, or artificial intelligence system without express permission. All rights reserved. Inquiries may be directed to Simon & Schuster, 1230 Avenue of the Americas, New York, NY 10020 or permissions@simonandschuster.com.

Copyright © 2025 by Nathalia Holt

All rights reserved, including the right to reproduce this book or portions thereof in any form whatsoever. For information, address Atria Books Subsidiary Rights Department, 1230 Avenue of the Americas, New York, NY 10020.

First One Signal Publishers/Atria Books hardcover edition July 2025

ONE SIGNAL PUBLISHERS / ATRIA BOOKS and colophon are trademarks of Simon & Schuster, LLC

Simon & Schuster strongly believes in freedom of expression and stands against censorship in all its forms. For more information, visit BooksBelong.com.

For information about special discounts for bulk purchases, please contact Simon & Schuster Special Sales at 1-866-506-1949 or business@simonandschuster.com.

The Simon & Schuster Speakers Bureau can bring authors to your live event. For more information, or to book an event, contact the Simon & Schuster Speakers Bureau at 1-866-248-3049 or visit our website at www.simonspeakers.com.

Interior design by Davina Mock-Maniscalco

Manufactured in the United States of America

1 3 5 7 9 10 8 6 4 2

Library of Congress Cataloging-in-Publication Data has been applied for.

ISBN 978-1-6680-2774-5
ISBN 978-1-6680-2776-9 (ebook)

For Larkin

There are no words that can tell the hidden spirit of the wilderness, that can reveal its mystery, its melancholy, and its charm.
—Theodore Roosevelt
African Game Trails, 1910

CONTENTS

PROLOGUE	The Last Large Mammal	xi
CHAPTER 1	The Happy Valley	1
CHAPTER 2	The Valley of Death	19
CHAPTER 3	The Crim's Pool	30
CHAPTER 4	Eaves of the World	39
CHAPTER 5	House of the Prince	55
CHAPTER 6	South of the Clouds	67
CHAPTER 7	Forge of Arrows	89
CHAPTER 8	Complete Heaven	102
CHAPTER 9	Kingdom of the Golden Monkey	118
CHAPTER 10	Temple of Hell	132
CHAPTER 11	Land of the Yi	150
CHAPTER 12	The Hall of Asian Mammals	167
CHAPTER 13	The Summer White House	186
EPILOGUE	Trail's End	201
ACKNOWLEDGMENTS		209
NOTES		211
INDEX		245

PROLOGUE

THE LAST LARGE MAMMAL

Two brothers smoothed a map on the table in front of them. The land they were examining was colored in greens, browns, and grays. Running across the map, like the stripes of a tiger, were irregular white blotches. Each blank space represented the unknown, a section of the map still unplotted and unexplored. The squiggly dotted line of a river, UNKNOWN printed in small text, cut through the white. It was 1928 and the world was still a checkerboard of wonder, the continents imperfectly mapped.

Ted and Kermit Roosevelt, the two eldest sons of former president Theodore Roosevelt, were planning an adventure. Although they consulted maps from a diverse range of cartographers, including those drawn in China, the unexplored regions persisted. The vast Asian continent dappled with white spoke to them.

The world was full of explorers, all examining maps like the ones the Roosevelts possessed. There was a heady, optimistic feeling that persisted among them. No one could be certain which mountain was the tallest on earth nor which trench in the ocean the deepest. Every expedition held the possibility of making its members world-famous explorers.

PROLOGUE: THE LAST LARGE MAMMAL

Ted and Kermit Roosevelt, 1926, courtesy of the Library of Congress.

The 1920s were a decade of discovery, as groups of scientists, adventurers, and hunters ventured forth into the wilderness to fill museum collections. They were successful: every large mammal on earth had been attained, and their bodies mounted in exhibits, except for one.

The Roosevelts desired this one animal so acutely that they could barely speak about it with each other, much less anyone else. "We did not let even our close friends know," wrote Ted of their shared purpose. Some dreams sound too wild when spoken aloud. The animal the Roosevelt brothers coveted looked like no other species in the world. It was a black-and-white bear so rare that many people did not believe it was real. This legendary creature was called the giant panda. Rumors swirled about the mysterious animal. No one, not even naturalists who had worked in China all their lives, could say precisely where the creature lived, what it ate, or how it behaved.

Brown, black, and polar bears had never been in doubt among

PROLOGUE: THE LAST LARGE MAMMAL

humans. Even polar bears, although living in the remote reaches of the Arctic, were well known, and had been kept in zoos for thousands of years. In Egypt, King Ptolemy II had a polar bear in his zoo in Alexandria as early as 285 BC. In 1252, a polar bear was part of the Tower of London's extensive menagerie of beasts.

Yet the same could not be said of the panda bear. Even among those living in the Republic of China, spanning some 7.7 percent of the earth's landmass, few had ever caught sight of the creature. Dozens of names were used to describe what might be a panda. In different dialects they called it "spotted bear," "giant bear cat," "white bear," and "bamboo bear," although no one could be sure that all these different names were referring to the same species. There were probable references to the giant panda in Chinese literature as early as the third century, although the descriptions were mythical, describing yellow-and-black creatures that munched on copper and iron. "While there are tantalizing stories implying that one Chinese emperor or another knew all about panda," wrote one author, "there's one great mystery. Why is there not a single rendition of this endearing beast in any of imperial China's illustrated natural histories?"

For Westerners, only the pelt of the panda offered proof of its existence. While stationed in the remote mountains near Chengdu in 1869, a French missionary named Armand David hired a group of hunters and asked them to go into the wilderness to collect interesting specimens. One man returned with the lifeless body of a strange animal, small in size, seemingly a cub, but with a striking black-and-white coat unlike anything they'd ever seen before. David inspected the animal closely and then skinned it, shipping the unusual pelt all the way to Paris to be examined by experts.

David gave the peculiar beast the scientific name *Ursus melanoleucus*, translating to "black-and-white bear." It wasn't a name that would stick. Nothing about David's finding was particularly scientific. The hide he'd sent might be stark in its coloration and completely different from any other specimen the museum in Paris possessed, but that did not mean scientists were ready to believe that this odd creature was real. A skin without a skeleton could be anything. Certainly, they did not intend to name a new species that, for all they knew, only a handful of Chinese hunters had ever seen alive. All they had was a vague description of a black-and-white bear and a small pelt of fur stored away in a museum.

Nineteenth-century illustration of the giant panda by Alphonse Milne-Edwards.

To this point in time unnamed, the elusive creature became known as the panda, yet even this label remains mysterious. It arose sometime around 1870, supposedly of French origin, although this has been debated. Some researchers trace its roots to the Nepali words for the red panda, *nigalya ponya*. While unrelated biologically, the red panda was well known in Southeast Asia, and first described by Westerners in 1825.

In 1916, a party of German explorers was traveling in China and Tibet. They sought the legendary black-and-white bear, but no matter how hard they looked, they could find no evidence of the creature. In the small village of Lianghoku, their leader, the German scientist Walther Stozner, asked a group of local hunters to bring them pandas, dead or alive. According to the explorers' written account, the hunters returned with several dead pandas, along with one very young panda cub. The Germans tried to feed the infant milk, then a flour slurry, and finally sought out a human wet nurse, but their efforts all failed. A few days later, in the hands of the first Westerner to glimpse a live panda, it died. No specimens were brought home, however, so the skepticism continued.

Then, in 1919, a missionary named Joseph Milner made an unusual donation to the American Museum of Natural History in New York City. "A giant panda," read the announcement in the museum journal *Natural History*, "from eastern Tibet, one of the rarest of animals.... The striking black and white coat, short muzzle, and curious black patches about the eyes gives it a very extraordinary appearance. Almost nothing is known of the animal's habits." Milner—simply a buyer in this case—himself had nothing else to contribute about the artifact, but even the skin of the panda was enough to excite scientists. Here was tangible evidence that this animal existed.

Expeditions were immediately launched to find the panda. Everyone knew that whoever got the panda first would enjoy immediate fame, both in scientific circles and in the press, so the number of expeditions venturing to China skyrocketed. But even the most experienced explorers came home empty-handed. As the years went by, cynicism grew, leaving many to suspect that the creature was not real at all, merely a black-and-white phantom sent to confuse and plague its seekers.

Ten years passed and the legend of the mythical panda continued to grow. Documenting the animal appeared hopeless to most explorers. The objective was not simply to shoot the bear, although even that would be welcome at this point, but to return with a detailed description of its habitat and diet so that the animal could be unveiled to the public, its existence proven beyond a doubt. Ted and Kermit, experienced in hunting but novices in science, seemed highly unlikely to stun the world with their findings.

The American Museum of Natural History was well known for sending adventurers out into the unknown. Ted and Kermit's connections with the museum ran deep: their grandfather was a cofounder of the institution and their famous father, Theodore Roosevelt, had helped fill its halls with lions, tigers, elephants, and bears. Even Kermit had contributed. In 1909, as a college student, he had accompanied his father

PROLOGUE: THE LAST LARGE MAMMAL

on an expedition to Africa. The elephant calf he shot on that trip stood in the center of the Hall of African Mammals, surrounded by seven adults, all poised as if mid-stampede.

Although the calf weighed over two hundred pounds while alive, he looked small compared to the members of his family. While the grown elephants often scared visiting children, with their massive bodies, sharp tusks, and raised trunks, the calf always drew a crowd of young admirers, some of whom would look into the elephant's glass eyes and dream of becoming explorers and scientists themselves one day.

When Kermit looked at the animal, he was transported back to that moment in time, when he was just nineteen years old and desperate to gain his father's praise. His father had explained to him that this was not a mere hunting trip but a scientific expedition. "I can be condemned," the former president argued, "only if the existence of the National Museum, the American Museum of Natural History, and all similar zoological institutions are to be condemned."

The idea of killing animals in the name of conservation might now seem bizarre and even cruel, but Roosevelt's philosophy was not unusual. Naturalists of the era purposely killed endangered species, preferably an entire family of them, in the name of science. Even Charles Darwin, the father of evolutionary biology, was a skilled hunter who used his assortment of rifles to collect specimens. In Uruguay in 1833, the famed naturalist shot several species of deer that would, in later years, become endangered.

For those hunting animals on the brink of extinction, the rationale was clear: if a species was doomed, there was no point in protecting the last few remaining individuals. Nature had made her choice. Instead, scientists should learn as much about the species as possible. This sadly meant killing the animal to study it scientifically. While today human intervention can protect some, although not all, endangered species, it is only successful because of the foundational work scientists have performed over the centuries. Without understanding the basic biology of the species, scientists would not have the means to save them.

PROLOGUE: THE LAST LARGE MAMMAL

The American Museum of Natural History agreed, writing, "Should some interesting mammal disappear from the face of the Earth before such a permanent, concrete record of it could be prepared and stored up for posterity, museums would have indeed been derelict in their duty." What they didn't mention, of course, was that a museum fills its halls and pays its bills not with complex scientific names but with people clamoring for strange and exotic beasts.

By the 1920s, the museum in New York City had funded expeditions to every continent, and its halls were filled with a diverse range of specimens. Most of its specimens, however, would be hidden from the public—roughly 90 percent of museum collections remain in storage. Yet surveys could enrich the scientific community, adding to the field's knowledge of biodiversity, evolution, and conservation. Descriptive field journals would offer explanations of the species explorers encountered, but physical specimens were more valuable, offering proof of their discoveries and baseline raw morphological data (such as the curve of the mandible and the length of the spine) to demonstrate evolutionary change. These collections would become vital to the future of conservation biology.

Although the Roosevelt name could be found throughout the halls of the New York museum, the venerable institution wasn't sponsoring Ted and Kermit's expedition. Neither brother had proven himself as a scientist or explorer. Instead, Chicago's Field Museum was taking a chance on the Roosevelts. The expedition would be funded by Illinois businessman William Vallandigham Kelley. Museums depended on wealthy donors, whose money bought them halls and exhibits bearing their names, even though they never walked a step of a trail themselves.

The wealthiest Americans, the Rockefellers, Carnegies, Astors, Whitneys, and Vanderbilts, were swimming in money in the 1920s. Stocks had quadrupled in value, to peaks never seen before, and the market seemed destined to go up, up, up. Philanthropists were eager to fund expeditions likely to result in popular exhibits, where their name would be touted by the press and admired on brass plates.

During America's Gilded Age, these prominent families had amassed an unimaginable level of wealth. The richest man in the world, John D. Rockefeller, was worth some $400 billion in today's money. Similarly, J. P. Morgan had so much cash that he was called upon, not once but twice, to bail out the federal government. It would take a century, not until today's era of tech billionaires, before the country would once again see such wealth concentrated among such a small number of Americans. Wealth inequality today is peaking, with the proportion of assets now held by the top 0.1 percent of the population identical to that of the Roaring Twenties.

"Neither Kermit nor I can afford this on our own," Ted admitted as they plotted their course. The brothers' fortunes were not as plentiful as many supposed. Their father, Theodore Roosevelt, had inherited $60,000 from his father in 1878. Less than a decade later, Roosevelt lost most of his fortune in a risky cattle ranching investment out west. By the time of his death in 1919, his wealth was concentrated in the eighty acres he owned on Long Island. He'd been slowly selling off chunks of the property to raise funds. The land and most of his fortune were left to his wife, Edith, while his five children split a trust fund of $60,000, worth $1.5 million today. Ted and Kermit inherited $12,000 each, a massive sum at the time, but not quite enough to make them independently wealthy. However, they did have some experience in the field, and not only alongside their father's larger-than-life persona.

In 1925, the Field Museum had funded the brothers' expedition to Central Asia. At a time when Percy Fawcett was disappearing into the Amazon and Roald Amundsen was leading the first air expedition to the North Pole, the wanderings of President Roosevelt's eldest sons in India and Pakistan in search of an elusive sheep had not caught the imagination or attention of the press. Instead, it was scientists who took notice.

It wasn't just a sheep that the Roosevelt brothers had found, but a legendary bighorn. Called *Ovis poli* (or *Ovis ammon polii*), it had been described by Marco Polo in 1256. This was the first time the animal was displayed in an American museum. When the Field Museum first received the pelt, skull, and precise measurements of the *Ovis poli*, the scientists

PROLOGUE: THE LAST LARGE MAMMAL

were in shock. It had been so long since anyone had seen the sheep that the species was thought to be extinct. However, it wasn't long before *Ovis poli* began drawing attention in other circles. The animal's dramatic spiraling horns, unique to the species, were just the thing to tempt hunters. The brothers had no idea that their actions would lead to the sheep being hunted to near oblivion.

Now the Roosevelt brothers were after the last large mammal unknown to science, and they believed that if they were willing to push farther, deep into the Himalayas, the ultimate prize would await them. "Central Asia . . . is the mecca of our desires," wrote Ted. However, as with the *Ovis poli,* the unintended consequences of their journey were not yet clear. What was certain was that no one would emerge from the wilderness unchanged, and one of them wouldn't return home at all.

Once on the ship, the first leg of the journey underway, Kermit felt a familiar feeling of trepidation fill his gut. "It's a warning," he wrote, "that something unexpected is about to happen."

CHAPTER 1

THE HAPPY VALLEY

A trail once traced an ancient path across the largest continent on earth. It started on the coast of the Indian Ocean, before edging into the damp rainforests of Myanmar. It followed a fast-flowing river, its banks teeming with life, then rose high on a ridge so narrow that a mule's hooves inevitably slipped in the dust, kicking the air perilously above a two-thousand-foot precipice.

Not all the explorers along the trail were human. A small seed caught in the crack of a tree. Its roots wrapped around the trunk and then began to stretch across the forest, its reach infinite. It was a banyan tree, and its aerial root system followed the trail as if it too knew the way.

The trail crossed China and Tibet. Humid forests gave way to vast, windblown savannahs, desolate and forbidding. The gusts that rushed down the trail cried to its travelers to turn back, as the danger was increasing. The peaks of the Himalayas rose above the plains, the snow, ice, and wind a permanent, deadly fixture no matter the season. There was no tent strong enough to withstand the mountain squalls and no fire hot enough to warm the explorers' hands and feet. Instead, the cold closed in, stealing away consciousness at the roof of the world.

There were animals and plants in these wilds that no other human being had ever documented. Birds appeared that looked as if sculpted from silver, every feather lying perfectly in place. Strange aquatic creatures, taller than a human from end to end, splashed in muddy pools. Monkeys with long, golden tufts of fur and bright-blue faces hung from the trees. Yet the most surprising animal of all lay deep within the heart of the trail, inside a hidden kingdom whose entrance required months of hard trekking.

No living person can now tread the Roosevelts' path. The trail they walked has been wiped clear, its roads paved, and many of its forests decimated. The people who once found sanctuary in its walls of green are long gone, along with many of the species they encountered along the way. Still, in the quiet of the wilderness their legacy remains, a murmur in a preserved bamboo forest.

THE SOUND OF THE FOREST at night is different from the serene hum of daytime. The darkness heightens the senses so that the shrill call of an owl or the chirp of the cricket is amplified. Every growl from the brush echoes with ominous undertones. Every rustle of the leaves has the potential to shake the confidence of even a skilled explorer. The expedition had just begun, only a single day spent on the trail, but already the group was groping in the darkness. A scientist, one of their own, had vanished.

Tai Jack Young looked down the darkening trail and felt a spurt of fear. Herbert Stevens, an English naturalist, had stepped off the trail six hours ago, and never returned. The Chinese wilderness, usually teeming with life, had gone quiet around him. There was no telling where Herbert might be, or even if he was still alive.

They were hiking from Burma into China on a path known as "the back door," so named because of its remote and rugged entry point into China. There was no border guard to greet them and no sign to mark the way. Instead, the trail merely narrowed slightly. The explorers would not even have known that they had entered China had Jack not told them.

Tai Jack Young went by "Jack" for the convenience of English speakers. His last name had already been changed, from Yang to Young, by his grandfather, Young Tak Cho, who felt the name rolled off the tongue of Americans better, closer to the true Chinese pronunciation. He was just nineteen years old and devastatingly handsome, tall with thick hair that he liked to comb back from his face into a pompadour. He loved wearing crisp shirts under modern, three-piece suits. Now, however, his hair was disheveled and his clothes dirty.

He'd been hired as the expedition's interpreter and guide. Jack was the youngster of the group, although in many ways more experienced than his employers. He knew China thoroughly, thanks to his upbringing in a small village outside of Hong Kong and his travels as a child. His father had been born in San Francisco in the United States, his mother in Guangdong, a coastal province in Southern China, but he was from neither country. His birthplace was Kona in Hawaii, an island territory positioned between both nations, reflecting his own fractured identity.

Now Jack was on the trail and eager to prove himself. He admired Kermit Roosevelt's relaxed attitude in the woods. In New York City, Jack had been lured by the brothers' prestige and famous name. Here in the wilderness, the younger brother had a different draw. Kermit was comfortable in his skin, more fully himself while trotting the dirt trails than he had been on the concrete sidewalks. Kermit was a man who had traveled across the world, from Africa to South America to Asia, and Jack, just on the verge of manhood, couldn't help but be awed by his accomplishments. What he hadn't yet glimpsed were Kermit's weaknesses.

Jack had met Kermit earlier that year—1928—when New York City was abuzz with news that the two eldest sons of former president Theodore Roosevelt were undertaking a new expedition. As soon as the trip was announced, men and women of all ages, including an entire Boy Scout troop, began writing letters to Ted and Kermit asking to join:

"I am eighteen. I have always wanted to see the world and this is my opportunity."

"Well, Colonel, when do we start?"

"I would like to go with you on one of your expeditions, and I saw in the papers that you are going on another one. . . . I am 11 years old and my chief occupation is going to school gee its awful."

Instead of applying directly to the two Roosevelt sons, Ted and Kermit, Jack had approached the Chinese embassy, where he worked part-time, and explained the situation to his boss. He was a journalism student at New York University and had no connections, but he did have one skill they needed. He was adept at languages and knew multiple Chinese dialects—and, what's more, had traveled through Southwest China as a child with his father. He was young and had never served as a guide on an expedition or as a scientist on a field mission, but he knew if someone like the Roosevelts gave him the opportunity, he could prove himself.

"He was a slight, nice-looking boy," Ted wrote, after they decided to hire Jack. They immediately sent him to the Field Museum in Chicago so that he could begin an "intensive course" in the scientific techniques of specimen collection. The curator, however, was skeptical. "Doubt if Chinaman can be trained in two weeks to be of value to the expedition," he wired to the Roosevelts.

Jack felt the weight of expectations as night approached. The group was on the verge of descending into panic. "We have only an hour until it's dark," Kermit explained to his brother, Jack, and Suydam Cutting, another naturalist on the expedition. "We'll have to split up into search parties around the site where Herbert left the trail. Be back at camp by nightfall. We can't risk losing more of us out here in the dark." They agreed, and Ted and Suydam veered to the left while Kermit and Jack took the right. It was a strange, reckless sensation to lift one's boot off the safety of the dirt trail and plunge it into the green of lush vegetation, but they had no choice. They had to find Herbert.

by the brothers' glamour. With his gun slung over his shoulder, Kermit held a compass pressed tightly in his fist. Although the light was fading, Kermit was doing his best to keep track of their position on the map.

They soon came upon a mossy green embankment and Jack could feel his boots slip dangerously in the loose dirt as they hiked down. The trees were so dense in this part of the forest that they first heard water before they spotted it. It started as a low rushing growl and then grew progressively louder, eventually filling Jack's ears as they approached.

"This isn't the Taping River," Kermit yelled over the noise of the water. "It must be a tributary." Jack wasn't sure—the stream seemed wide enough to be a river—but he nodded in agreement. Water lapped against the banks and swirled around the rocky streambed. The trees had thinned along this section of the forest, and the open sky above the water was like a funnel for the last gasps of daylight. Kermit and Jack scanned the water hastily, but it was Jack's young eyes that spotted it, a small raft, just a speck of black on the edge of the blue water.

Kermit called out again, this time as loud as he could, but he was too far for the sound to carry over the rush of the water. "Let's chase it!" he yelled.

They ran down the streambank. It wasn't as easy as it seemed. Even though the path wasn't crowded like the jungle's interior, there were still tree roots rising from the mud to trip them and thick, chest-high brush that ripped at their clothing to slow them down. Kermit yelled again, and this time, the men on the raft looked up.

Jack felt some trepidation about calling out to the strangers; after all, they had no idea who they were, and here in this remote jungle, it could be just about anyone. When he was twelve, he'd traveled through this region with his father. That trip had taught Jack—unlike the Roosevelts, who moved through the world with blind self-confidence—the dangers that lurked in the unknown. Jack and his father had been robbed in a remote area like this one, and while fortunately unhurt, he'd learned to be wary.

As Jack assessed the situation, he saw there were two men aboard, both standing upright. One of the men held a long bamboo pole in his hands and was using it to steady the raft near the shore. Their figures were indistinct

From left to right, Suydam Cutting, Theodore Roosevelt Jr., and George Cherrie having breakfast on the hunt for *Ovis poli*, 1925. Courtesy Field Museum, CSZ51815.

Jack didn't know Herbert well; they'd only met a few weeks earlier. Herbert Stevens was a biologist originally from the University of Cambridge who now lived in India. As a scientist, Herbert was invaluable. He was an expert in zoology and botany, able to identify a wide range of species, from birds to insects to trees. However, his reputation was lacking within the scientific community. The Royal Geographical Society had rejected him as a fellow, stating his lack of fieldwork and publications. Jack suspected the real reason was that he preferred to live abroad, rather than moving within London's social circles. Similar to Jack's motivations, this expedition gave Herbert an opportunity to prove that the revered geographical society was wrong about him.

While Kermit called out Herbert's name, Jack looked back toward the trail. He couldn't spot its flat contours, and for a moment he felt a surge of panic. The prospect of failure—becoming lost during this trip, or losing a member of the party—had not crossed Jack's mind until now. These were the Roosevelts. They bore an air of invulnerability that had carried the entire group forward into this treacherous environment. Even Jack, one of the few who understood what he was getting into, had been blinded in the dusk. The sky had gone a deep indigo, which reflected in the water and cast a blue glow across everyone in its path.

"Herbert!" Kermit screamed as he and Jack finally approached close enough to distinguish faces. They had finally found their wayward scientist—but who was this mysterious man with him?

"Climb aboard," Herbert told them, and to Jack's surprise, Kermit did. It was a strange follow-up to a rescue, but Jack had no choice but to hop onto the wobbly bamboo raft as best he could. It was crowded with all four of them. As Kermit began chatting with Herbert, Jack introduced himself to the man holding the long pole.

"My name is Saw Bwa Fang Tao," the man said, bowing his head in greeting. Jack returned the custom and then looked at him curiously. The Shan people, like the Chinese, list their surnames first. Jack scanned his memory, certain he had heard the name Saw Bwa before. Then it came to him. Two nights prior he had heard it mentioned around a campfire. This was no mere fisherman they had come across—it was the brother of Saw Bwa Fang Yu-chi, the ruler of the Shan state.

Fang explained that he had crossed paths with Herbert in the woods and immediately sensed that the scientist was lost and in need of help. He had decided to take Herbert to the closest town in hopes of reuniting him with his party. Now he could take all three of them upstream. Jack couldn't help but shake his head at Herbert's good fortune. Thanks to Fang's care, Herbert would still have made his way back to camp even if they hadn't found him.

Jack took up another long bamboo pole that was strapped to the raft and helped navigate as they moved on the water. He was happy that they wouldn't have to scramble back up the steep gorge, but soon his misgivings about climbing onto the raft proved prescient. Just three hundred yards upstream, the water became so shallow that they had to take to the shore, dragging the raft with them. Once on the water again, the topography of the stream tricked them, and they fell over a rocky drop, upturning the raft and drenching the group in muddy, cold water. They had no choice but to flip the raft over, grab the poles, and try again. Now that the panic of finding Herbert was over, exhaustion set in. The physical toil of pulling the raft became excruciating.

The dark didn't help. Night had fallen and the sky was inky black overhead. The clouds had moved to obscure the stars, and the moon drifting between them cast a frosty December glow. Jack and Fang navigated the raft toward the shallows where the town of Kanai met the water. Jack was the first to get off, because someone had to get their feet wet, and he pulled the raft close to shore. He didn't mind the water seeping into his boots, since he was already thoroughly wet and muddy at this point. As he pulled, he noticed that Kermit and Fang were engaged in conversation. Fang was bowing his head slightly and pointing to the far shore.

"He's inviting us to breakfast tomorrow morning," Kermit explained as he stepped off the raft, "and I've told him we'd be honored to accept."

"Do you know where we're staying?" Jack asked.

"Everyone knows," Fang replied amid being pulled away by the stream.

Jack woke up early the next morning, expecting to be one of the first out of bed, and was surprised to see Ted and Kermit packing up already. "We barely slept," they explained as they shook out their bedrolls, "it was so loud." Jack had slept outside in one of the tents and couldn't imagine what noise the two men would find to bother them inside the small country inn. He had considered them fortunate to have beds to sleep on, while the rest of the party had dozed in tents on the hard ground. When he peeked into their room, however, he immediately understood. Two chickens had sneaked into their quarters and were now perched atop the beds, squawking loudly as if annoyed to find their room occupied by intruders. Jack laughed at the sight, which prompted a sharp look from Ted, who didn't find the circumstances amusing.

Lack of sleep made the Roosevelt brothers grumpy, and Jack listened patiently as they complained about the delay to their plans by the necessity of having breakfast with Fang.

"If it wasn't for him, who knows where Herbert would be," Suydam reminded them, and Ted and Kermit nodded grudgingly as they continued packing.

Suydam Cutting was about their age, thirty-nine, and tall, over six feet, with dark wavy hair and an irresistible smile. Although relatively inexperienced as a scientist, he radiated confidence. It was impossible not to listen to him.

"Yes, yes, but let's get it done early," Kermit said as he stuffed his possessions in his rucksack. Then he turned to Jack and told him the plan for the day. Jack was the one who would then explain to the local guides who accompanied them that their party would be divided; the guides should go ahead without them, and Ted, Kermit, Herbert, Suydam, and Jack himself would catch up with them by the end of the day.

The past three years had taught Suydam that he cared little for taking the lead. His ego was not fragile; he never demanded credit or desired fame. Instead, his pleasure came from the immediacy of travel and adventure. "He is a man of undaunted courage," noted Ted.

Suydam Cutting. Courtesy Smithsonian Institution Archives. Image SIA2008-0775.

Suydam had been born Charles Suydam Cutting on an icy winter day, January 17, 1889, in New York City. He entered this world in a large, sprawling penthouse on Manhattan's Upper East Side that overlooked Central Park. While the Roosevelt brothers ran wild in the forest that bordered their home, Suydam knew only the tame green paths of a park crowded with people.

Early on, it seemed unlikely that Suydam would ever be tempted by the wild parts of the globe. He stayed within the paved lanes expected of him, graduating with a degree in engineering from Harvard University in 1912 before starting work in sales. A stint in the United States Army during World War I brought him briefly overseas, but he soon returned to society life in New York City. Then, in 1925, he bumped into his friend Kermit Roosevelt traveling on a train from Boston to New York.

As soon as Kermit spotted Suydam, he strode to his row and plopped down beside him. "By the way," Kermit said casually, as if the pair were already in the middle of a conversation, "we have an expedition planned." Suydam politely smiled, greeted his friend, and then asked the requisite questions: "Where to?" and "Who's going?"

It was nothing but a pleasant chat between two friends until Kermit began describing the Vale of Kashmir, a valley formed from the foothills of the Himalayas in Central Asia. He told of villages that few Americans had ever seen, treks through thick jungles, and climbs up dusty mountain passes that would conceal strange and exotic animals, including the rare blue sheep. It reminded Suydam of Marco Polo's romanticized tales of travels in the region in the thirteenth century. Then, out of nowhere, Kermit invited Suydam to join them. They would set sail in two short months.

The offer was impetuous and Suydam was hardly qualified. He was no scientist, nor was he a big game hunter or even an experienced traveler. There was no earthly reason for Kermit to invite him. And there was certainly no motive for Suydam to accept. He had a steady, if dull job, but with a riotous social life in a city he adored. Yet despite all the reasons he should tell Kermit no, Suydam blurted out an enthusiastic "Yes!" His eagerness shocked even himself. For the next three years he kept saying yes, tagging along on all sorts of expeditions with a range of different naturalists, from Chinese Turkestan to Assam to Ethiopia.

There was something about the Roosevelt brothers and their expeditions that drew people to them. "They were everything to me," Jack would later say. Every wish he had for his future was intertwined in Ted and Kermit's larger-than-life personas. Yet he did not yet realize how fragile the brothers truly were. At thirty-nine years old, Kermit seemed self-assured in the woods but was timid in nearly every other aspect of his life. He wasn't a politician like his father, nor a prosperous businessman like his brother Ted. Even his marriage was faltering. Only in the forest did he feel like himself.

Ted, although Jack saw him as the confident elder brother, was secretly miserable. He was forty-one years old and for the past seven years had pursued a career in politics and failed. Where his father had a wild exuberance and a rugged, cowboy persona, Ted was different. Although also a war hero, having volunteered among the first soldiers to go to the Western Front and fought in major battles during WWI, he was not nearly as boisterous or loud, even when he tried out the family's trademark shout "Bully!" while campaigning.

While working as the assistant secretary of the navy, Ted had leased oil fields on public lands to private corporations. These leases, made without competitive bidding and found to involve bribery, are now better known as the Teapot Dome scandal, after the Wyoming oil field of their extraction. One of the contracts in question would end up in the hands of Ted's younger brother Archie, vice president of the Union Petroleum Company, a subsidiary of the company that had won one of the oil contracts. Through Ted was personally found innocent of corruption in the matter, he was "politically obliterated," in the words of his wife, Eleanor.

The Chinese wilderness was camouflaging the two Roosevelt brothers, and Jack saw them only as twin suns, brilliant and bright.

THE BEAST IN THE CLOUDS

THE EARLY MORNING LIGHT WAS throwing mist around the surrounding hillsides as Ted, Kermit, Suydam, Herbert, and Jack left the inn. Fang was waiting for them in the street, and he greeted them with a booming "Hello!" that made the travelers smile. He was the kind of man who was impossible not to like, gracious, friendly, and endlessly chatty. As they walked through the town, he spoke of the area's history, giving an abbreviated tour of Kanai as they strolled. Unlike with Jack, to whom he'd given his name freely, he'd introduced himself to the others as Philip Tao, his English name. He didn't trust their tongues to pronounce his name properly.

"Market square," he explained, as they passed a dusty courtyard, mud-packed walls lining its boundaries, and empty except for a few wooden animal stalls.

"How do the goods travel here?" Kermit asked. It was a reasonable question. The trail they had hiked on was narrow and treacherous and the stream they had poled last night was far too shallow to accommodate a large boat. It was hard to imagine how farmers got their wares to the stalls.

"Backs or mules," Fang explained. He then asked the Roosevelts how the dirt roads could best be widened for automobiles. He told them that he was thinking of bringing a car to the area in pieces, hauling it by mules over the mountains and then assembling it right here in Kanai. It was obvious he had put extensive thought into plans for the town and had a keen eye for improvement.

"Is your brother thinking of improving the roads?" Kermit asked, but Fang shook his head sadly. There was obviously more to the story but no one wanted to press for details.

The travelers followed Fang across the water, hopping from rock to rock across the stream they had floated on last night. It seemed shallower in the morning light and less treacherous than it had the night before. On the other side, a hillside rose green and lush before them, and they walked up a small path to a house that was built atop it.

Kermit began to admire the gardens. They were terraced up the hillside and filled with plants and flowers he'd never seen before. There was so much beauty contained within them that Herbert began to linger, oblivious to the others, and Ted had to physically tug on his arm every so often to keep the scientist apace with them.

Fang didn't mind. He was proud of his gardens and loved to show them off. There were acres of fruit trees and tiny lakes that dotted the green landscape like drops of morning dew. Fang pointed to the closest of the small lakes, where a tiny island glimmered in the morning light. "That's where we'll have breakfast," he explained.

The group toured the house, far larger than any other in town, with two stories and a spacious floor plan. They asked about his family, and Fang spoke of his three brothers: one was young and did not speak English, another was twenty and living in Japan, and the third was the eldest and the hereditary ruler of the region. "Is he here? Will we meet him?" Ted asked. Fang shook his head. The silence lingered uncomfortably.

The group walked down the hillside and climbed aboard a bamboo raft. Kermit laughed, remembering how muddy they had been the night before, and they poled over to the island, this time without mishap. A small rustic house sat there, with a roof and no walls, but it was comfortable and had thick cushions on the floor with a view out over the lake and the rice paddies beyond. Intrigued by the birds on the water, Herbert asked about them as he watched ducks and geese fly overhead in a tight V before circling down toward the water.

The Roosevelts began talking about their expedition, the animals they hoped to find, and the exhibits they wanted to bring to the Field Museum in Chicago. As they spoke of the grandeur of the museum and the many animals from around the world it already contained, Jack was watching the Roosevelt brothers curiously. It was clear from their manner that Ted and Kermit thought they were impressing Fang. Instead, their host found them foolish and wasn't afraid to say so. "Why spend so much money," the brother of the ruler asked, "only to get very tired?"

Breakfast was delivered to the island. The men sat around a charcoal fire on top of which was placed an earthenware pot. Meat and vegetables bubbled inside, and the explorers were encouraged to add greens and condiments that sat in jars in front of them.

Kermit, Ted, Jack, and Suydam dove into the food, but Herbert couldn't seem to get a bite in. He had little experience using chopsticks, so the utensils slipped through his fingers. The tighter he squeezed them, the more apt they were to go flying across his plate or fall into his lap. A ripple of embarrassment passed over the explorers. Herbert, their poor wayward scientist, always seemed to be getting into trouble, whether in the jungle or outside it. And their expedition no longer seemed so worldly and competent. Nonetheless, Fang smiled indulgently, as if accustomed to guests who were unable to make do when asked to venture beyond their Western comfort zones.

Kermit plucked a rooster's head from the pot and good-naturedly plopped it into Suydam's bowl. Jack tossed a handful of herbs in the pot, and Ted sat contentedly, looking out over the water. The group was enjoying a relaxed and happy meal until Ted asked, "Where's your brother? Is he joining us?" Fang's previous reaction had given no indication he'd be comfortable with this kind of personal inquiry about his elder brother, but they plowed ahead anyway with all the confidence of men accustomed to getting the answers they desired.

"Have you heard of the poppy? Opium?" he asked. Everyone around the table nodded.

Fang began to tell them of his brother Yu-chi. He had once been a powerful man, the ruler of the Shans, a position that passed down from father to eldest son. But then everything changed. He tried opium and quickly became addicted to the drug. He now spent his days mostly unconscious, languishing in grimy opium dens far from home. It was impossible to reason with him or get him to return to his responsibilities, although Fang tried. Yu-chi had long given up any pretense of ruling the region, and although it was still his name that echoed across the valley, it was his younger brother who took charge of the actual management.

Addiction drains life wherever it can, regardless of wealth and power. Fang explained to them that they would see opium addiction everywhere in China, from large cities in the valley to small towns in the mountains, like this one. The drug, he said, was omnipresent. It was likely, he said, that they would soon encounter bandits, whose movements followed opium traffic across the mountains. There was grief in his words, a constant sadness when he spoke of his brother, the Shan ruler who had possessed every advantage but lost it all in the euphoric rush of poppy seeds.

Opium had taken away more than beloved family members; it had stolen the peace of China. The drug was the direct cause of civil strife, corruption, and military tyranny in the region as it created a vast pool of wealth that men and women used to wage war and buy influence. The poppy was woven into the fabric of China's economy, and its grip on politics was inescapable.

"Those poor souls," Ted replied with anguish, sympathizing with Fang and the plight of those under the drug's influence; yet what he could not comprehend was the culpability of Westerners in the rise of opium. While poppies had grown in China for centuries, and were used in medicine sparingly, the rise of the drug in China had not occurred until the eighteenth century when the British-held East India Company began to import Indian opium into the country. The British had become obsessed with Chinese tea, and a vibrant trade ensued between the countries. Soon millions of pounds of the drug were being consumed in China every year.

Attempts by the Chinese government to prohibit the United Kingdom from importing the addictive drug were met with the full might of the British armed forces in the first of the Opium Wars in 1839. Thoroughly dominated by the British, China was powerless to protect its citizens; opium imports rose tenfold, and Hong Kong was ceded to the foreigners. By 1906, roughly 25 percent of the population smoked the drug. "Never before or since," read one report, "had the world known a drug problem of this scale and intensity." By the time the Chinese government was finally able to ban the drug in 1906 it was too late; opium had taken hold of the country.

The spread of opium had coincided with violence and war across China. Fang told the explorers how he had awoken one morning to find soldiers in the village, tearing through homes and businesses. The people immediately took to the hills where they began raining down stones on the soldiers below. In the tumult, Fang's mother was taken captive. "They put her in a cage," Fang explained, voice trembling with emotion.

"A cage?" Kermit repeated, and turned to Jack. He couldn't believe that such an account was true. He wanted Jack to confirm that nothing had been lost in translation. Jack did so, although he wasn't shocked. He knew what atrocities the National Revolutionary Army was capable of, and they were far worse than the Roosevelts could imagine.

When Jack was four his family had moved from Hawaii to Southern China, and by the 1920s were running a prosperous automobile shop, one of very few able to service American Fords and Chryslers. Then, just six years later, they lost everything when the Kuomintang nationalist party's forces under a new commander in chief, Chiang Kai-shek, swept through the area. He directed his soldiers to rout out opposing leaders in armed conflict and then take whatever bounty they wanted from the shops they passed by. Jack's family was left with nothing.

Sensitive to the Shan people's losses, Jack sympathized with Fang. Kermit asked to speak with the mother, to express their deep condolences for all the family had endured, but after recovering she had left China for Japan, to live with her other son.

Indeed, the only other member of the family who seemed to be present among the gardens, lakes, and trees was Fang's baby boy. He was just one year old, and the unofficial Shan leader presented him to the group proudly. His legs and arms were chubby with baby fat and he gurgled happily while Kermit cradled him in his arms. Kermit was entranced. Yet as he admired the baby, Kermit realized how lonely Fang was. He was isolated here, with only his wife and son, and missed his large, vibrant family. The effects of war and opium combined had stolen so much from him.

"Will you pick an English name for him?" Fang asked Kermit, noticing how enamored the Roosevelt brother was with his son. He asked for one that began with a *W* and Ted and Kermit discussed it earnestly. It was hard to think of *W* names, and they finally settled on Walter, which Kermit thought easy to pronounce and spell.

"Walter," Fang repeated as he took back his son and cradled the child in his lap. The baby held all the wishes for his future and now he had an English name chosen for him by the sons of an American president. To Fang this would be a welcome omen, especially for a family in dire need of good fortune. He insisted that they accept a small gray pony as a token of his gratitude.

It was impossible not to think of the de facto Shan ruler and his baby as they hiked out of the happy valley, the gray pony trailing behind. They climbed a steep hill, and as they reached the peak, they saw endless terraces of rice paddies in concentric circles beneath them. They looked like shards of glass, cut into the earth of the hills and sparkling in the afternoon sun. Hundreds of cranes collected on their green and brown edges and Herbert was ready to go wander into the wild again, but the four other men restrained him, promising that their path would soon curve down into the terraces, and they'd get a closer look without needing to leave the safety of the trail.

Herbert's pace quickened. He loved birds of all kinds, but cranes were special to him. There was a beauty and an elegance to the birds that concealed the sheer power they contained in their wings. The cranes below them were able to splash in the rice paddies here near the border with Burma and India and then spread their wings and fly all the way over the Himalayas, more than a thousand miles distant. Cranes had long been symbolic across cultures in Asia, where they represented happiness and eternal youth.

No matter the reason, Kermit was content to see Herbert moving faster. They still had to catch up to the rest of their party, who were waiting for them on the trail ahead. Once they reached the top of the pass, it began to decline steeply, the forest closing in on them again and obscuring the view of the rice paddies in curtains of green. When they emerged through the thick trees, they found themselves on the edge of a paddy, with hundreds of cranes gathered just off the trail.

"Put your guns down!" Herbert yelled when he saw how close the cranes were. "Down on the trail, immediately!" It was counter to everything that Ted and Kermit had been taught by their father, but they complied with the scientist's request and dropped their weapons on the ground. "Follow me," Herbert commanded as he led them off the path so they could get a closer look. It was strange for Herbert to take charge like this and his unexpectedly authoritative tone made everyone immediately obey.

The group edged closer to the birds, maintaining a respectful distance and careful to be as quiet as possible. Herbert told them that they were sarus cranes, the tallest flying bird in the world. There was a grace to their movement; merely standing in a rice paddy they looked like avian ballet dancers waltzing in the water. "*Antigone antigone*," Herbert explained, writing out their scientific name in his field book as he sketched the birds. He explained that because of their bare, long necks they had been named after Antigone, the daughter of Oedipus, who had hung herself.

In Greek mythology, Antigone's tale is a tragic one, and Herbert worried that the cranes were headed for a similar fate. He noticed how their presence was intimately linked with the wildness of their habitat. "The terraced slopes of rice cultivation," Herbert wrote in his field journal, "cannot support them."

It was just a few lines scribbled in a plain brown notebook, but it was the first time that a scientist had labeled the crane as an indicator species, calling attention to the threats posed by deteriorating water conditions and a dangerous lack of wetlands. As the group of explorers looked out over the fields of rice cultivation, tens of thousands of sarus cranes roamed the wilderness below them. In a mere generation, fewer than a hundred would remain. When Herbert urged the group to put down their arms, he may have saved a few lives, but it was hardly enough.

Soon the five men would walk back to the trail and the explorers would pick up their guns again. They were ready to kill something.

CHAPTER 2

THE VALLEY OF DEATH

Kermit held the .22-gauge rifle firmly against his shoulder. He was crouched low to the ground, keeping his breathing even and inaudible so that all he could hear was the chirrup of birdsong emanating from the forest around him. He had just seen something—a quick flash of brown amid the green of the jungle. Kermit's rifle was absolutely the wrong gun for hunting in the forest, lacking the power to take down large mammals. As annoyed as he was by the weapon, he had no time to swap it out. Instead, he remained as still as the trees around him and wondered what he had just seen moving through the brush.

Kermit's eyes caught on an animal. It looked like an otter below him. He blinked, unsure of himself. There were no reports of wild otters in this part of China. Still, there it was, a bundle of brown fluffy fur at the water's edge. It moved awkwardly on the shore the way aquatic mammals tend to do, bumping across the land in bursts like a baby learning to crawl. Kermit knew he had to act quickly. Once the otter reached the water its movements would turn agile and sure. He would never catch it then. It was possible that he was looking at a new species, one that could lend insight into the wilderness that surrounded him. Like the cranes he had viewed the other day, river otters are an indicator species; their absence is a sign that waterways have become dangerously polluted.

In North America, Kermit knew, river otters were disappearing fast. It was a topic his father had spoken of. In some states, especially those where the human population was booming, there were hardly any to be found. Yet hunters kept looking. Their thick, waterproof fur was so highly prized that the effort of trapping the remaining animals was worthwhile to most fur traders.

Kermit was searching too, although he would never classify himself alongside the fur traders. He felt a thrill run through him as he considered what to do next. His descriptions would mean nothing to the scientific community, and a picture could only capture so much. There was only one path ahead and it required the rifle in his hands. He took aim and squeezed the trigger.

It was an action he had done thousands of times before, a movement so practiced that he sometimes woke up with his pointer finger wrapped around his blankets, shooting animals in his sleep. This time, as he squeezed the trigger, he couldn't see the creature fall, but he had a gut feeling, a sort of animal instinct, that he'd hit his target. The bullet had split the silence, and now Kermit scrambled to his feet noisily as he plunged into the bushes. The otter was gone but a spot of blood could be seen on a green leaf nearby. His instinct had been right.

Kermit immediately followed the animal into the brush. It was wounded and needed to be put out of its misery. This was the reason he didn't like to shoot with a .22. His father had taught him to never leave a creature in pain, especially when his shooting was the cause of it. He yelled to Jack that he would be back and began to trail the animal. He was already feeling regret. There wasn't time for a prolonged chase; they had to get back on the trail. He knew everyone was waiting for him, and after Herbert's misadventure, patience was thin for those who wandered off trail. Kermit ran as far as he dared into the woods but then stopped suddenly in frustration. An animal was hurt, perhaps even dead, and it was all for nothing.

THE VALLEY OF DEATH

Kermit was quiet as he reached the dirt path where the others were waiting. He held the rifle in his hands for a moment. Like his father, he wasn't a particularly good shot. He'd never excelled at marksmanship, although his dad had assured him that it wasn't important. "Perseverance, skill in tracking, quick vision, endurance, stamina, and a cool head," comprised the elder Roosevelt's hunting advice. As Kermit held the weapon, a gun that had taken a shot but hadn't yet taken a life, he began to feel uneasy. He wondered if he was capable of this.

"When Father went off into the wilds," Kermit wrote, "he was apt to be worried until he had done something which would justify the expedition and relieve it from the danger of being a fiasco." Kermit was similarly seized by feelings of inadequacy and fear. They had been on the trail for a little more than a week but hadn't collected anything for the museum. Doubt had crept into Kermit's mind and was beginning to multiply. The panda had always felt like a far-flung dream, but now he worried that they would return with nothing of interest. It would be the complete humiliation their father had always feared. With each day that left them empty-handed, they were approaching disaster. So many explorers had failed on this quest before them that it was hard to keep the faith.

The group was climbing a long, sleep stope that would soon drop precipitously into a region known as the Valley of Death. As they passed through small towns, many people warned about what lay ahead. "You don't want to go that way," one woman told Jack. "It's full of evil spirits that haunt people while they sleep."

"What do you mean, haunt people?" Jack had asked, a chill running up his spine.

"They go to sleep and don't wake up."

Jack knew it was just a ghost story, but he still found it unsettling. Even if it wasn't evil spirits killing people, it was clear they were heading into an area where many people had died. Yet even if they wanted to, they couldn't avoid the Valley of Death. The trail wound right through these hills, sinking down to a sparkling green river that was obscured by thick green vegetation.

As the trail wound deeper into the jungle, Kermit was reminded of his father's last expedition. In 1913, Kermit and his father had traveled through South America with a large party led by Brazilian explorer Colonel Cândido Rondon. They had survived the trip, although just barely. The former president had emerged from the jungle frail, having lost fifty-five pounds, a quarter of his bodyweight, in three months. They had mapped the River of Doubt, later renamed Roosevelt River, but the fifty-six-year-old explorer would never undertake another expedition. "The Brazilian wilderness stole ten years of my life," Roosevelt professed. He died five years later, in 1919.

An overwhelming feeling of claustrophobia seized Kermit as the forest walls rose on all sides. It was like being encased in a room of green. The trees were so thick and tall that even the sky was blocked from view. Suddenly he was plunged into the past, remembering what it felt like to reach his absolute physical limit: the deep pit of starvation that gnawed at his belly, the ache of his limbs as he forced himself to paddle upstream, and the utter helplessness of watching his father fade away, tormented by fever. The one thing he was certain of was that expeditions were unpredictable. Sensing his friend's uneasiness, Suydam turned to Kermit. "It's too oppressive," Suydam complained of the jungle around them, but Kermit only nodded back.

Sweat gathered on the explorers' necks as they rose in elevation. The trail was so rugged that the sides of the mules' bodies pulled in fiercely as they struggled to breathe. Kermit's feet pounded the dirt, but he still gripped the gun in his hands, not ready to stow it in its carrying case on a nearby mule. They'd been warned several times of bandits who roamed these hills, and it seemed prudent to hold on to a gun just in case. He might not be able to shoot an otter, but he could at least scare off an outlaw.

Kermit and the other men made up the back half of their traveling expedition as they scrambled over the rocky, narrow trail. At the front was a clutch of women. "You can't imagine what power these women have," Ted wrote in a letter to his wife after the trip. "Never underrate them."

THE VALLEY OF DEATH

The women he spoke of were guides, those individuals with superior knowledge of the mountains and superlative endurance.

Such guides were sometimes called porters, and other times coolies, a word thought to have derived from the Hindu term *kuli*, or person who carries baggage. In the nineteenth century, *coolie* became a racially charged term to describe indentured laborers working abroad and was used widely as a derogatory term for anyone of Asian descent.

"The entire Chinese coolie class," said President Theodore Roosevelt in his 1905 State of the Union address, "that is, the class of Chinese laborers, skilled and unskilled, legitimately come under the head of undesirable immigrants to this country, because of their numbers, the low wages for which they work, and their low standard of living. Not only is it to the interest of this country to keep them out, but the Chinese authorities do not desire that they should be admitted."

No matter what belittling language was used, guides had always played a key role on expeditions. They were invaluable, often walking twice the distance of the explorers they worked for. They ran ahead on the trail, making sure the route was clear, and then prepared the campsite for the night, pitching tents and readying scientific equipment.

While a handful of men would take all the credit, it took a team of men and women to load and carry the equipment, guide the mule team, and make camp. The women on the expedition were carrying packs laden with gear, guns, and food. "More than half our porters were girls and women," wrote Ted.

The packs the region's guides carried were sometimes impossibly heavy, "all the way up to four hundred pounds," Ted noted. Such a heavy burden was more typically carried by tea porters, who hoisted packages of precious tea leaves over a thousand miles, from Chengdu to Tibet, where the drink was beloved; because China controlled its exports tightly, tea could be conveyed no other way but on the backs of men and women.

While the tea porters suffered under their massively heavy loads, the guides on the Roosevelt expedition carried only a fraction of this weight, roughly fifty pounds; the mules could carry the rest. Their worth was measured in their ability to guide rather than their capability to transport. Each guide also carried a "stout stick" on which they could rest the weight of their pack when they wearied.

The women's knowledge and athleticism were evident in defined calf muscles, superior fortitude, and extraordinary skill in navigation. They were excellent with maps, and deeply informed on the environs. They easily estimated distances and planned routes that stretched upward of a thousand miles. Trained in mountaineering as well, they were able to belay their companions, cross dangerous crevasses, and, most of all, survive the Himalayas, where even in summer the highest peaks on the planet were covered in snow.

Their inner strength corresponded to a great force of will. In casual conversations, translated by Jack, Ted and Kermit were struck by these women's strong opinions. They pushed the team to cover greater distances on the trail and were constantly on the alert for danger. As Jack chatted with them, he learned that many had multiple husbands, most commonly three, although one woman had seven at home. While the job of a guide was not easy, it drew numerous applicants. A woman could make a small fortune, nearly five times the annual income in rural China, in less than a year. In addition, there was a notable respect for those who accompanied Western expeditions. Like the Sherpa people in nearby Nepal, the position allowed them to display their skill and stamina.

Western China was, Ted declared, "the land of women's emancipation." Describing the strength and power of the women who worked for him, Ted wrote, "The custom of polyandry is largely, if not entirely, responsible for the amelioration of women's lot. Unattractive and distasteful as it may seem to an occidental mind, there is here a great deal to be said for it." As impressed as Ted was with the women guiding the party, he was about to become awestruck.

Suydam and Ted were toward the front of the party, behind the advance group of women porters, while Kermit and Jack stuck to the back of the pack. Bringing up the rear, as usual, was Herbert. The other men grumbled about him, but Herbert couldn't help himself. With his binoculars bouncing around his neck and his eyes searching the trees, he was constantly being tempted off trail by the promise of new and interesting species. "He's a nuisance," Ted remarked quietly to Kermit. "We need to talk to him again."

After his disappearance the first day on the trail, he was trying to keep up, but when he saw a streak of vivid yellow-green, he stepped off the trail, a net clutched in his fist. A butterfly was sitting perfectly still on a branch and Herbert gazed at the insect in wonder. He could tell the butterfly was a swallowtail by the shape of its wings, but the species was unknown to him. The hind wings glowed an iridescent green that was so ferociously bright it looked electric.

The color was brilliant, almost too beautiful to be real, and in a way it wasn't. The color didn't exist, at least not in the traditional sense. Unlike the turacoverdin pigment in many bird feathers or the melanin pigment of human eyes, there was no green pigment contained in the wings. Instead, the green was the result of an unusual three-dimensional structure. The wings contained layers of spiral-shaped cells that acted like a glass prism, changing the direction and speed of the light, so that human observers witnessed a shimmering green. When the butterfly's wings flapped, the angle of light shifted and so did the color, creating flashes of light that were meant to confuse a predator. It was an illusion of geometry, a protective nanostructure that Herbert might not fully understand but could definitely appreciate. It was an insect that lived only in this small slice of the world, and he had been lucky enough to find it. With one swipe of his net he lunged for the butterfly . . . and missed.

Kermit looked on at Herbert in frustration. It was afternoon and they were making steady progress on the trail, but no headway in building the Field Museum's collection. He had already missed the otter this morning, and now they couldn't even catch a butterfly.

Most scientific papers of the era reported the features of the natural world as a dry set of facts. They read like a grocery list of mandibles, egg sacs, and feathers. Herbert's papers were different. They covered a range of species, from birds to mammals to insects, revealing an unusually broad interest and experience. Even more uncommon was the content. Herbert used his papers to advocate for greater protection of natural habitats. "Wherever private enterprise has safe-guarded its interests by conserving even a tithe of the indigenous forest," Herbert wrote in 1923, "this has all been in favour of the birds, and can well be appreciated by all true lovers of nature's marvelous and bounteous gifts. Where no check has been kept on the primitive methods of land devastation . . . the ultimate issue has been disastrous in many respects."

Because so many animals depend on insect populations, Herbert was becoming more interested in how their presence, and especially their absence, shaped how ecosystems function. The butterfly Herbert was chasing might be pretty, but just like the otter Kermit had wounded, it could also tell them something about the health of the jungle they hiked through.

Many scientists who studied bugs in the 1920s were focused on "economic entomology," or the study of pest species that prey on human food crops. But a growing movement was emerging that focused on the importance of the insects themselves, and the need to catalogue and preserve them. These scientists called themselves "heroic entomologists"; many of them were amateurs, untrained in the field but beginning to understand how fragile the population truly was. Herbert was intent on contributing to this field and becoming "heroic," as it connected his interests in scientific study and environmental preservation. Unfortunately, despite his good intentions, Herbert was a magnet for bad luck.

On an expedition to Vietnam in 1924 he'd collected a vast number of insects, birds, and small mammals for the British Museum. It hadn't been easy; the area was unknown to scientists and poorly mapped. But he'd found some unusual specimens, perhaps even a few new species. As he considered what he would name the animals, he thought of his wife, Amy. The scientific name *amyae*, he thought, had a nice ring to it. Amy had wanted to join her husband, but Herbert told her it was too dangerous. Amy was like him, unafraid of collecting insects, even the poisonous ones, and willing to hike miles to reach wilderness unsoiled by human civilization. Had she been born in a different place or time, perhaps in Western China, she might have accompanied him.

His field notebooks were full of drawings, detailed notes, and maps of the area he'd traveled. He'd brought his camera along on the expedition, to document every creature that he could not kill. On his way back, he was traveling by boat down the Srepok River when disaster struck: the steamboat capsized in deep water. Caught under the ship, Herbert fought for his life, eventually pulled from the depths cold and nearly lifeless. The bulk of the specimens he had collected so carefully, along with his notes, camera, and film, were all gone, and Herbert was lucky he wasn't at the bottom of the river with them. So much had been lost, and for a while Herbert wallowed in misery, unsure if he could ever explore again. Now fate was giving him another chance. If he was lucky, he might even find a new creature to name after his precious Amy, cementing the link between his love and his work.

There was a human connection that existed between the trees the explorers passed, the insects that swarmed them at night, the frogs that sang on the slow-moving edges of the rivers, and the mammals that they eyed greedily through the scopes of their rifles. With every footstep the explorers took, every note they scrawled into their field journals, they were endangering that delicate balance of life, even if they couldn't yet appreciate the effects their journey would have. Instead, they plodded on quietly, ruminating on their failures and wondering if the day could possibly get worse.

The group of women ahead led the party onward. When the visibility stretched for miles, the guides sang songs to each other and swapped stories. Yet when they were rounding steep curves in the mountains, or the green of the forest crowded their vision, they turned quiet and wary. While Jack was hearing rumors of the deadly spirits that inhabited the Valley of Death, the women were worried about thieves. The guides knew their expedition was vulnerable, and unlike the men who hiked behind them, they carried no guns.

The women hiked alongside the laboring mules, loaded down with camping equipment. The trail was steep and dusty. As they approached the rise of a hill, the figures of several men suddenly appeared in front of them. The women backed up a few steps, tentatively assessing the situation, when they heard footfalls behind them. They whipped their heads around and realized with horror that they were surrounded. The men inched closer, tightening the circle, and pointing their guns threateningly.

If they did not act, the women knew they'd be dead. The thieves who roamed this remote hillside were rumored to leave behind a trail of corpses. It wasn't just their lives in jeopardy; if they failed to defeat these men, the rest of the party would likely be taken by surprise.

With one swift movement, a woman slapped the rear of the nearest pack mule as hard as she could, sending the animal bolting through the circle of outlaws. In the chaos of the moment, the guides charged their adversaries. The bandits weren't expecting such action and quickly fled, determined to at least nab the runaway mule for their efforts. The guides wouldn't allow it. They knew that if they let the thieves leave, it would not be long before their party was again surrounded. One woman turned back to inform Ted, Kermit, Jack, Suydam, and Herbert of the encounter, while the rest of the guides scrambled up the hillside in pursuit of the thieves.

When the women caught up with the miscreants, they beat them with their stout sticks and their fists, leaving the men broken and bloody. A few outlaws sneaked off into the hills, but the guides were satisfied that they were no longer a threat to their party. They had proved that their expedition wasn't to be trifled with, and figured that any remaining menace would disappear. The women grabbed the mule and led it back to the trail. Sweat was dripping down their faces, flushed from the effort, as they rushed back. They were eager to lose as little time from the encounter as possible, especially since they still had to reach their destination and set up camp for the men. They had dealt with the crisis so quickly that Ted and Kermit could hardly comprehend the danger they had just evaded.

The Roosevelt brothers did not appreciate the perils that surrounded them in the Valley of Death. China was in the midst of civil war, one that had started in 1927, in which the Kuomingtang nationalist government and the Chinese Communist Party were in direct conflict. With armed battle ongoing, the violence was spilling out into the countryside. Soldiers who had deserted the army were gathering in small groups along trails like this one, eager for weary travelers who carried valuable loads. It was an exceedingly dangerous time for those moving across Western China.

The Roosevelts were blinded by their position in society. Educated in Ivy League schools and possessing all the advantages and connections of their family, they moved through the world with singular confidence. Yet having a famous father could not save them here, or even open their eyes to their considerable vulnerabilities. Traveling in a small group without military protection, they had no idea what they were getting into.

As the group descended into the Valley of Death, the rumors that Jack had heard of evil spirits lurking in this fertile land were about to come true. The bandits had been a warning from the universe perhaps, one that the explorers had not heeded. Now, as the trail transformed from steep mountain passes to lush green jungle, one of their own was about to experience how dangerous the valley truly was, and this time no woman would be there to save him.

CHAPTER 3

THE CRIM'S POOL

Kermit felt his intestines twisting themselves into knots. The pain was unbearable. He couldn't think, couldn't talk, could barely even summon the energy to breathe. He lay curled in the fetal position, remaining as still as possible. His brother Ted sat next to him and every so often placed a hand on his wrist to feel his pulse. Like his breathing, it was erratic.

Ted was no doctor, and he had no way of knowing what was wrong with Kermit or how serious his condition might be. While his hands rubbed his brother's back lightly, Ted's mind raced with the possibilities. They had been hiking every day for the past two weeks, sleeping on the ground in their bedrolls or atop stacks of hay in barns infested with vermin. They'd been eating all sorts of unusual food, and while they often boiled their water before consuming it, they sometimes drank directly from streams out of desperation.

The past few days had been particularly difficult. The trail had followed the Taping River, whose waters were clear and cool. It was nearly impossible to resist the temptation to cup the cold water in their palms and drink deeply. The deeper they traveled into the jungle, the tighter the valley walls closed in on them. It reminded Kermit of South America in an unsettling way, a feeling that something fatal was lurking in this Valley of Death. The Roosevelt brothers had been dismissive of old superstitions, but now that Kermit was ill, Suydam's apprehensions, when he complained about a "sense of oppression," seemed prophetic.

Ted felt increasing panic seize his body. His bond with Kermit was special. Their family was large, comprising six siblings, but Ted and Kermit as the eldest brothers and only two years apart in age were exceptionally close. "Neither Kermit nor I really enjoy fully exploring or hunting unless we are together," Ted wrote to a cousin. The sentiment was true: in each other's company, the two men were happiest and most themselves. Ted knew how much his younger brother depended on him. As children, Kermit had rarely left his side, always wanting to play with his big brother. Adulthood had not stopped Kermit's dependence on Ted, merely changed the nature of it. They had exchanged their toy cap guns for real rifles, and now the familiar woods of Long Island were instead the vast jungles of the Himalayas, but Kermit's yearning for his brother's company had not dissipated.

Now Ted felt helpless. He had no means to communicate his suffering. He did not know the words for *doctor*, *medicine*, or even *help*. The illness had revealed how comprehensively the remote wilderness had removed the practiced ease he had become attuned to. As Kermit sank lower, in the grips of such misery that every member of the expedition began to fear for the man's life, Ted took Jack Young aside. "Something needs to be done," he said in a low tone. Jack nodded in agreement. It was painful to sit and watch Kermit writhe in pain. He would much rather be looking for help and was only waiting for Ted's permission to do so.

They were camping on the outskirts of a village, so he set off to find someone, anyone, who could help Kermit. He was acutely aware that it was up to him, the youngest member of the expedition, to save the Roosevelt brother. With dark thoughts racing through his mind, he eagerly approached the first man he came across.

"Hello," he said in Cantonese, but there was no response. Then the stranger began talking in an eager, excited manner, but Jack couldn't understand a word. He wasn't ready to give up; after all, Kermit's life depended on him. He tried every dialect of Chinese he knew, then moved on to Arabic and Hindustani. Nothing worked. Out of desperation, Jack tried French, but that was no better. He moved on sadly, determined to find someone who would understand the problem and direct him to medical help.

The encounter was a reminder of how frustrating the trip had been thus far for Jack. A day earlier he had attempted to speak to four men whom the group came across in the bazaar of the village near which they were camped. The Roosevelt brothers had wished to talk to them, drawn in by their distinctive felt boots, heavy coats, and knives hung somewhat menacingly from their belts. No matter what Jack tried, however, he couldn't make himself understood. It was infuriating to speak so many languages and yet to find himself unable to make basic conversation. Jack knew the men were from Tibet, whose border lay many miles west of their location, but they might have been from the moon for all the communication he was capable of.

Jack felt himself caught between worlds and his identity had never felt so fractured. Ted called Jack a "Chinaman," even though he'd been born in Hawaii, the son of an American father and a Chinese mother. Yet here in this remote section of China he was considered a Westerner, one who was struggling to communicate with the villagers he met along the way. On this expedition he was learning a painful truth: he was not American enough for the Americans or Chinese enough for the Chinese. Jack felt adrift, as if he belonged nowhere.

The desire to succeed in this endeavor was central to Jack's ambitions. He needed to impress the Roosevelt brothers with his skill in languages and especially his ability in scientific research so that they'd recommend him to colleges in the United States. His craving was for education, even though he wasn't sure where it would all lead. No matter where he studied in the US he would be a "Chinaman," and the prejudices he experienced because of his background and the color of his skin would persist.

Despite his inner torment, Jack knew that Kermit's life rested on his ability to find a doctor immediately. It was the middle of the night, in a strange part of the world, and with tales of marauders everywhere, but Jack worried that if he didn't find someone soon, tragedy would strike. There was a rumor in the village of a doctor who had studied at a hospital in Rangoon, and Jack was ready to knock on every door until he found him.

Hours later, Jack located the doctor, dragged him out of bed, and brought him to Kermit's bedside. The doctor examined Kermit quietly, giving no opinion or diagnosis, not even to Jack who was standing beside him, ready to translate. Instead, the doctor pulled out a small tin box painted in bright colors. Inside were pills. He handed Kermit the unidentified medicine and then, without a word of instruction, stood quietly by. No one thought to question the doctor, not even Jack who could communicate with the man. They were simply too worried to waste time on inquiry. In complete ignorance of his illness and its treatment, Kermit took what was handed to him and meekly swallowed the medicine.

Whether the mysterious pills helped or it was merely the passage of time, Kermit woke the next morning shaky, but better. He pulled on his boots with a ragged determination. He would not be the cause of the party's failure. Ted was adamant that he preferred waiting for Kermit to improve, but his brother shook his head. It was time to march.

It was market day, and the empty streets through which Jack had run the night before in search of help were now filled with stalls. Duck, the meat dried and flattened in circles, was hung from rope where it swayed in the breeze like wind chimes. Kermit entered an apothecary stall where rows and rows of porcelain jars were on display, filled with powders and liquids. A row of deer horns stood prominently on a shelf, but Kermit didn't know what they were for. As he searched the labels of the many jars, he was sure there was some medicine there that would be useful to him, but without Jack or one of the guides, he was helpless.

Kermit's eyes continued to search the shelves when an overwhelming smell overtook him. He whirled around and noticed slimy strips of entrails hanging from the ceiling. It seemed impossible that he hadn't seen them right away. Flies buzzed madly around, swarming the raw meat. Kermit felt his stomach lurch; it was empty and sensitive from his sickness, and a wave of nausea overwhelmed him.

A whirl of people flew across Kermit's vision, his head was spinning, and he was about to faint. Ted wasn't there to catch him. His brother loved markets and had already moved on to the next row of stalls, happily enthralled with the many wares before him. Jack noticed Kermit, however, and rushed to his side, bracing himself against his employer while Kermit dropped his head into his hands and waited for the light-headedness to pass. It was a long pause before he could straighten up and continue.

After the urge to retch had subsided, Kermit watched the people moving around the marketplace. Everyone was there to sell or shop, driven by a purpose, but Kermit had his own resolve. His brother was easily distracted by trinkets, always had been, but Kermit was looking for something important. A shortcut.

From across the market, Kermit saw monkeys chained to a stall and made his way over to the animals. This seemed like the right section to investigate. Jack stayed by his side, worried about Kermit's health. Finally Kermit spied what he was searching for: a seller hawking fur pelts. He rummaged through the collection excitedly, his fingers flipping through the soft, fluffy skins as if they were pages of a book he was intent on reading. His eyes were searching for only two colors: black and white. Every other shade, even gray, was unwelcome. The vendor was talking to him, but Kermit couldn't understand what he was saying. His body ached from lethargy and his brain felt fuzzy. Fortunately, Jack was at his side.

"He's asking what you're looking for," Jack said.

"Tell him we need a panda."

"I did. He says that he has some. Look." Jack pointed to the seller's arms, but there was no telltale black-and-white fur contained in them. Instead, the pelts were a dark reddish color and small, about the size of a beaver.

"Ask for the black-and-white bear," Kermit urged, even though he wasn't sure he was looking a for a bear. He had no way of knowing what animal family the panda occupied. He might be looking for a giant cat or canine. Jack turned to the seller and began talking again while Kermit examined the red fur. Even though fur pelts naturally shrank after being tanned, there was no way this had come from a large mammal. It looked like it had been stolen off the back off a house cat. Kermit knew it must be *Ailurus fulgens*, also known as the red panda. The animal lived in this part of the world, Southwest China, India, and the Himalayas, and was not at all rare. The species had been described a hundred years earlier by Georges-Frédéric Cuvier, a French zoologist. He had chosen the genus *Ailurus* from the Greek word for *cat*, and *fulgens* from the Latin for *shiny*.

Cuvier had never traveled to China and knew little about fieldwork, but a preserved red panda had been brought to him in Paris where the scientist carefully measured the skin, jawbones, femur, and teeth before bestowing its scientific name, which the animal would keep, and classifying it phylogenetically within the raccoon family, a designation that would shift many times. Everyone knew Cuvier's name: he had coined the term *hereditary* and been inducted into Britain's Royal Society, and his name could be found throughout the Muséum National d'Histoire Naturelle in Paris, particularly the exhibits that displayed the red pandas as they were in life, poised on tree branches or curled together in a den as if sleeping.

Before the animal was brought to France, a team of explorers had traced the jungle paths. They'd found the red panda and recognized its red fur tinged with white on its face, black belly, and ringed tail. They then hunted and killed the animal, preserved its specimen, and shipped it to France. One of their number, the white man among them, would be cited in journals and papers, and then promptly forgotten. The rest of the team would never be recollected at all. Kermit knew that every step he took on this expedition was likely to be forgotten and every sacrifice he made unnoticed. But he was determined not to end up like the explorers who had come before him. The panda was special, he was sure of it.

"No," Jack said, "he doesn't have it, and hasn't ever seen fur like that."

Kermit nodded and turned away, depressed. Ted called Jack over to another furrier, where he was fruitlessly asking the same questions. They were seeking one of the rarest animals in the world, and they couldn't even be sure what colors they were looking for. Describing the panda with its bold coat of black and white was like describing a unicorn, an animal that lived only in stories, its appearance familiar, but without proof how could you ever know if it was real?

It was time to leave the market, and Kermit followed Ted and Jack through the crowded maze of streets. Kermit noticed a poster hanging on the side of a building. It was one of a series that China had produced to promote its slew of new laws. There were posters that decried the use and trafficking of opium, and others that denounced deforestation. This one showed a woman with her feet prominently stretched in front of her. NO FEET BINDING! it proclaimed in Mandarin, in letters as brightly colored as the smile on the woman's face. As Kermit stared at the poster, with Jack translating the words, a young woman walked in front of him, hobbling on her deformed feet.

Although foot-binding had been formally banned in China in 1912, evidence of the practice was everywhere, in young children as well as old women. "Lotus feet" were seen as desirable, with some parents believing that the tighter the foot-binding, the richer the husband the girl would one day gain.

Binding usually was done to girls before the age of seven. The feet were plunged into hot water and rubbed with oil and herbs, the toes then taped together. The foot was beaten and all the digits, except the big toe, were broken and bound flat against the sole of the foot, forming a triangle shape. The foot was then bent in half, with the sole touching the heel, and tied with a long silk ribbon, before being placed in "lotus shoes." These platform shoes were tiny, approximately three to four inches, with a pointy toe, and made the girls' feet look as if they belonged to dolls. Every step, especially that first year before the toes went numb, would be one of overwhelming agony. The process could never be reversed.

It's believed that foot-binding was popularized in the tenth century by a dancer named Yao Niang. She bound her feet tightly in order to dance

THE CRIM'S POOL

on her toes for the Emperor Li Yu. The practice quickly spread across the royal court, where it was considered the height of fashion, much like the corsets of Victorian England. Eventually foot-binding spread to rural areas where it served a dual purpose. It was viewed not only as an allure for potential husbands, but as a way to keep young girls from running and playing, and focus them on household tasks, such as spinning yarn and sewing. In isolated villages like this one, which were connected to the outside world only by weeks of hiking on rugged trails, bound feet meant that the women were physically trapped within the valley they grew up in, likely forever.

Despite being banned by the new Republic of China government in 1912, the millennium-old cultural practice was not so easily overturned. Campaigns against foot-binding grew in the 1920s, resulting in only a small percentage of women in urban centers continuing the practice. In rural areas, however, foot-binding continued well into the 1950s. While the rest of the explorers looked on in curiosity, Jack had grown up with it. His mother had bound feet. Still, he agreed with Ted, who called it a "foolish and barbarous custom." The elder Roosevelt smiled when he saw a group of young girls running freely. They were the lucky ones: the first girls in a thousand years to escape the painful curse of broken feet.

As the expedition left the village, they found the trail crowded because of the market day in town. Throngs of travelers moved around them, some hoisting massive, ten-pound solid cubes of pink salt mined from Tibet, an eleven-day journey north. As they walked, Ted and Kermit discussed a monster that, according to legend, lay just ahead of them on the trail.

It was called the Crim. The beast was said to hide in the deep pools of the Mekong River, lying in wait for unsuspecting passersby. Jack described the creature as "a gruesome monster, resembling the shape of a huge blanket." The Crim had no preference for humans or livestock, but once an animal was close, it would suddenly rise from the depths and swallow up its prey. They'd been told the monster could gulp up individuals and even whole caravans of people if it so chose.

The legend of the Crim was not confined to this part of the world. A few years earlier, while in India, Ted and Kermit had heard tales of the Crim's exploits. Even so many miles away, the stories had centered on the Mekong River, which they were now approaching, the third-longest river in Asia, extending three thousand miles from the Tibetan Plateau to the South China Sea. The monster's description had not changed, whether in India or China, and the repetition of the same monster story seemed to strengthen its notoriety for the brothers.

Kermit knew it wasn't real, merely a legend like the Kraken or Loch Ness monster, but as they hiked up a steep gorge, he couldn't help but peer down at the Mekong River below them. The water was a milky brown, nothing like the crystal waters of the Taping. The sides of the mountain were steep, and as they climbed up, they were leaving the thick green of the jungle behind and heading into rocky mountain passes. Ted would call it "one of the most desolate places of the earth."

There was little that was green, even when the trail dropped back down to the river. It was a pronounced change after the jungle, the fields of rice cultivation, and the string of villages they had so recently passed. Now there was only the monotony of the brown-gray dirt before them and the rocks on either side, broken up by a series of wooden bridges that crossed the Mekong back and forth as the trail gained elevation.

It might have been that he was still weak after his illness, or the repeated tales of the Crim that he kept hearing, but as Kermit hiked the trail, an eerie sensation swept over him. The trail was about to cross a deep pool. Here was where the mythical monster lay. As they made their way across, Kermit looked down into the shadowy depths. He could not see the bottom. Then, without warning, something flickered at the edge of the water. He stood still for a minute and stared. There was a massive creature down there and its body was silvery white. "It's a monster," Kermit said aloud.

CHAPTER 4

EAVES OF THE WORLD

The waters of the Mekong were deep and murky. The wind was playing with the edges of the river, causing waves to run across its surface. Standing on the edge of a shaky wood bridge, the explorers found it impossible to guess how far below them the depths reached. Herbert, Ted, and Kermit were all sure they had seen something down there, hiding in the shadows of the water. They were all thinking the impossible: it was the Crim, the legendary beast said to swallow up passersby. If the legend was true, the group should leave the bridge immediately and catch up with the rest of their crew. The female guides were already far ahead, as usual, but most of the expedition, including the mules, had also passed them. It was just Ted, Kermit, and Herbert who could not look away from the waters.

The expedition had been on the trail only three weeks, but Herbert was seen as an irritation by nearly everyone. The reason: "He's slow," complained Suydam. The British scientist loved to wander and probe, much to the annoyance of the rest of the party who were trying to keep up the pace. This time, however, Ted and Kermit didn't mind falling behind the others. They were about to uncover the mythical being that lay below them in the water.

As the animal breached the surface, Herbert began sketching. A fin appeared, and then a smooth round head. It was not a monster at all, but instead a shimmering dolphin, five hundred miles from its expected ocean habitat. Even in the shadowy water, Kermit could make out its telltale curved body and gray skin. The animal swam slowly around the pool, occasionally raising a rounded fin as if in greeting. Kermit blinked at it dumbfounded; its presence made no sense to him. It looked like its oceanic relatives, and for a moment he wondered if the creature could be lost. The dolphin seemed as out of place in the brown water as their own exploring party appeared in the dense Yunnan jungle.

Herbert asked for someone to hand him a water sample. He placed a drop on his finger and then licked his tongue to the muddy water as the others looked on in horror. Boiling water was tiresome, and the expedition often skipped it, even though they knew the rivers were full of disease-causing parasites. But even when thirst overwhelmed them, they weren't foolish enough to drink from a muddy pool like this. It was asking for trouble. Herbert didn't seem to notice their disgust. "Salty," he said, nodding his head and writing the word *Brackish* in his notebook.

The Roman naturalist Pliny the Elder viewed dolphins as a kind of monster. "In the seas," he wrote, "spread out as they are far and wide, forming an element at once so delicate and so vivifying, and receiving the generating principles from the regions of the air, as they are ever produced by Nature, many animals are to be found, and indeed, most of those that are of monstrous form."

Dolphins are of course one of the most intelligent animals on earth. Their raw brain size is slightly larger than that of humans, and their brain-to-body ratio is higher than that of the great apes. Their neocortex, a region of the brain found only in mammals and that is responsible for consciousness, is complex, conferring advanced problem-solving, social interaction, and language.

Over the course of history, the human brain has tripled in size, growing most rapidly during a period of dramatic climate change that took place eight hundred thousand years ago. Glaciers and massive ice sheets rumbled across the continents, shaping the topography of earth and wiping out a stunning 99 percent of all humans. The Neanderthals went extinct during this time, and it seemed that humans were next. Struggling in the intensely cold climate, barely more than a thousand breeding humans were left on the planet. The next hundred thousand years would be critical to the continuation of the species. Larger brains allowed early humans to develop new means of subsistence within their surroundings. As earth became increasingly unstable, the evolution of the human brain was essential to our ancestors' survival.

Dolphins can trace a similar path of brain evolution. Their large brains allowed them to adapt to an environment untenable for marine mammals, seeking out the deep pools of the Mekong River, when other animals, despite their need for new habitats, could not. The dolphins that survived became a new species, tied to the rivers as deeply as their cousins were linked to the ocean. They were called a euryhaline species because of their ability to live in waters of varying salinity.

The animal the Roosevelts had spied was *Orcaella brevirostris*, the Irrawaddy dolphin. Locals knew the Irrawaddy to be shy, and distinctive for the small leaps it loved to make in the water, propelling its entire body out of the river, opening its blowhole, and breathing fresh air into its lungs. Then it would dive deep, holding its breath as it searched for food. During the rainy season, the dolphins traveled the river, but now, in the dry season, they remained in deep, muddy pools like this one.

Watching the dolphin play in the Mekong River was a reminder of what a complex brain is capable of, and how it allows a species to survive in an environment hostile to mammals. Humans had experienced their own evolution. A larger brain was necessary for humans to adapt to rapid temperature change as the climate alternated between glacial periods and warming trends called interglacials. Humans, like all animals, need pressure to reach their potential. As Ted and Kermit watched the dolphin swim, a direct result of brain size and evolution, they had little idea how much the wilderness they hiked through would alter them, and just how desperately they needed to change.

That night, arriving in camp late, the scientists set their traps as usual. Herbert, Ted, Kermit, Jack, and Suydam unfolded long, silken nets with a mesh so fine and delicate that it became invisible once it was strung up in the trees. Not just any trap would do; the party needed to hunt birds in a manner that preserved their skins and allowed for minute observation. The mist net contained hidden pockets so that after a bird flew into it, it fell and became tangled in a pocket below, still alive and ready for the scientists to examine it. Every day they checked their nets with the eagerness of children peering into their stockings on Christmas morning. They never knew what gifts China would grant them.

Setting up camp on the trail. Photograph by Herbert Stevens, 1929.

As the days passed, the explorers had noticed the morning air was becoming crisp. It was mid-January and over the past few days they had hiked up from the subtropical jungle into the mountains. The temperatures were plummeting. In their net that morning, Suydam found a laughing thrush. It was stuck firmly in the net, unable to flutter its wings but still alive. Suydam held the small creature in his hands and studied it. It had thick, white plumage at the top of its head and a jet-black eye mask, like a cartoon burglar would wear. Suydam recognized the bird immediately. He'd heard its song, which sounded like peals of laughter, all over the gorge they were hiking through. With a quick, sharp motion, Suydam broke the bird's neck and slipped it out of the pocket. They would take measurements, draw pictures, and then skin the birds later. It was miserable to hike with bloodstained clothes and hands.

As Suydam held the bird's delicate heart in his hands he was reminded of an expedition just seven months earlier, on which he had been assisting the noted anthropologist John Henry "J. H." Hutton in studying the Naga people of Northeast India.

One evening Suydam had watched a young boy approach their camp. The boy's skin was pallid, and he was shaking. Underneath his coat he held an awkward package, the size of a basketball. One of Suydam's colleagues took the wrapped bundle from the child's trembling hands and slowly stripped off the pieces of fabric covering it. A glint of white could be seen as the layers unraveled. Suydam watched closely as the fabric fell to the floor. Even though he knew what lay inside the package, his eyes were glued to the bundle. The anticipation gnawed at his nerves. With a last tug of the fabric, Hutton held the trophy up in the dim light so that the group could admire it. It was a human skull.

Families bragged of their vast skull collections to Suydam and their party of scientists. "Row after row of grinning skulls," Suydam described, "set up on perches, each standing for a savage murder." The skulleries were a legacy, passed down from father to son, and conferred prestige. Many Europeans dismissed the Naga as "savages" unworthy of study or understanding, yet this team of scientists had chosen to immerse themselves in their culture. Their aim was to describe the ethnic minority groups living in this part of India whose culture and traditions were mostly unknown to the outside world.

Collection of human skulls in India. Photograph by Suydam Cutting, 1928.

Speaking with the Naga people, Suydam asked one of the hunters which skull was the most valuable, expecting him to point to the head of a great warrior defeated in combat. Instead, a small baby's skull was held up. "The reason," Suydam wrote, "obscure at first, becomes logical as the Nagas explain it. A warrior will, of course, never take the head of a fellow-villager. He must stalk his prey in foreign territory. This entails danger, but in the long run the man's head is the easiest to acquire because men are obliged to leave their villages in order to tend to their paddies.... A woman, however, usually remains in the village or its immediate surroundings. To get *her* head means danger indeed, since the hunter runs the risk of being trapped by others.... For a baby, the hunter must enter not only a village but usually a house. It is the last word in daring, and it is no wonder that a baby's head is the most highly prized of trophies."

It seemed that no matter what expedition Suydam was on, science required death. He moved on to the next net, where a black-and-brown bird, all muted colors, struggled in the mesh. He grabbed it and the animal fit perfectly in the palm of his hand. At first glance it seemed like nothing special, but when Herbert took a closer look, he became excited. It looked like a warbler with its compact body and small, rounded wings, but its coloring was odd. Herbert had never seen a bird like it, and he suspected it might be a new species. They broke the extraordinary bird's neck and began to measure and sketch its features.

The new unnamed creature seemed to represent the endless diversity of China's rich wilderness. No matter where they turned, they were met with unusual species. As they hiked, the jungle felt like a limitless bounty, filled with interesting and beautiful wildlife. Yet each one of their footsteps was making an imperceptible dent in the species around them, not yet felt, but that would one day echo as loudly as birdsong through the gorge they were exploring. They were documenting a vanishing world, even if they could not yet appreciate it.

Along with the smaller birds, Herbert had caught two bright-orange ducks, known as ruddy shelducks or *Tadorna ferruginea,* and a falcated teal, *Anas falcata,* a waterfowl with bright green plumage on its head. The bright colors attracted other members of the expedition, and as Herbert sketched the birds' bodies, measured their wingspans and noted their colors, the men around him began to look on hungrily. Their provisions were getting low, they hadn't come to a village in days, and these ducks held interest beyond their scientific value. When Herbert was done, he wasted no time in skinning the animals and plopping the ducks in boiling water, so that the meat could be saved for their dinner that night.

Herbert was a vital member of their scientific team and helpful in general to the expedition, given his experience, but his idiosyncrasies and penchant to wander had become an annoyance. Ted and Kermit had taken him aside the evening before and explained to him that while his skill and his attention to detail were appreciated, they could no longer stomach the risk he had become to the expedition at large. It was time to part ways.

At first Herbert was shocked; he knew that the Roosevelt brothers, along with the rest of the crew, were frustrated with him, and of course his disappearing act on the first day hadn't helped his reputation. However, he'd always assumed he was essential to their work here. To learn that he was, instead, seen as an accessory was at first too much to take. He argued with the brothers, until he began to realize the advantages he could gain from the new arrangement. Without the Roosevelts constantly prodding him to move faster, he could take as much time as he wanted, wandering as far into the forests as he pleased. It was a happy idea, and Herbert imagined himself free from time constraints and able to explore fully at his own pace.

Picture of the caravan, taken by Herbert from far behind on the trail. Photograph by Herbert Stevens, 1929.

One disadvantage that Herbert did not fully consider, however, was the increased risk. Without the expedition, it would be just him and one or two of the guides. All the work of setting up camp, cooking, and preparing specimens would be split among them, making the days long and tedious. The danger, too, was heightened. A week earlier they had learned from a local magistrate that a "band of eight hundred Tibetan marauders" was roaming the countryside. For Herbert there would be no group of women to prowl ahead, scaring off bandits and scoping out dangers on the trail.

Even as Herbert debated the situation, he knew he had little choice. He was not in charge of this expedition and ultimately would have to abide by the Roosevelt brothers' wishes. To avoid conflict, he finally agreed, and the three men planned for how long Herbert would explore separately, and where and when they would rendezvous. They decided to meet in Tatsienlu, or the Forge of Arrows, some six hundred miles distant on April 15. They marked the location and wrote down the date on their maps.

The next day, as the rest of the expedition did their usual morning chores—collecting the birds, loading their gear on the mules, and throwing on their backpacks—Herbert stayed behind. He was no longer in a rush. He finally had all the time he needed. He stood by the campsite, the fire from breakfast still smoldering, and watched them leave.

Free of Herbert's plodding ways, the expedition set out on the trail at a brisk pace, aware that they had many miles in altitude to climb that day. The lush jungle, dolphins in deep pools skirting the riverbanks, and even the humid air they had enjoyed that morning dropped far beneath them. It felt a world away. The trail kept climbing, and their mules trod slowly behind them, the rocky path testing their endurance.

As they crested a ridge, Kermit looked down into the valley. The shoulders of the mountains were carved into islands of green with seas of dead brown in between. It was as if a giant had violently gouged the trees from the land, leaving only stumps and dirt on the hillsides. The damage hadn't been caused by a monster, at least not a mythical one. This was a vast clear-cut, and the logs of the freshly destroyed forest had already

Jade Dragon Snow Mountain in the distance. Photograph by Herbert Stevens, 1929.

been transported elsewhere. It was a depressing sight for anyone, but especially for the members of an expedition that needed wilderness to explore.

Just as China had imposed restrictions on foot-binding, new laws now targeted destructive logging practices. Deforestation had run rampant in remote forests like this one, where there was little enforcement and the price of the lumber was worth the penalty of breaking the law. Kermit couldn't help but connect the two practices in his journal. "Is it coincidence," he wrote, "that these laws should be passed?" They had witnessed countless women hobbling around villages, the arches of their feet broken beyond repair. The forests were similarly hobbled, with the species of the jungle broken up by the swaths of devastation. Both practices inflicted a lifetime of devastation and they both needed to end for the future of humanity. As they continued up the trail, the explorers were like the dolphins of the Mekong, searching for the deep pools of forest that were left to them after a very dry year.

EAVES OF THE WORLD

As they marched, it felt like they were opening a door to a new season. Every step higher in elevation brought new signs of winter, with the temperature plunging degree by degree, and a freezing hail that picked up intensity until it was prickling their faces. When they passed eight thousand feet, winter was suddenly everywhere, from the snow-covered peaks they could now see in the distance to icy rivulets that crossed the trail. Their bodies grew warm as they hiked and the cool air refreshed them, but every time they stopped, the sweat would evaporate on their backs and shivers would run up their spines.

There were no villages near and no trees to shelter under, so the guides chose a bivouac on a rocky outcrop. The winds whipped through the camp with such force that the edges of the tents thumped with rage. Outside, snow pelted everything in sight, and where once a breathtaking vista of the Jade Dragon Snow Mountain had appeared, now there was nothing but white in every direction.

Ted and Kermit sat huddled together at the periphery of their tent and began skinning birds. It was the last thing they wanted to do. Their muscles ached from climbing in the mountains, and their bodies were numb with cold. Yet they forced themselves, taking their sharp silver blades and running them along the birds' bellies in one swift, practiced stroke. Careful not to damage the brightly colored plumage, they gutted the animals, and prepared them for the museum.

Ted wished he could wait until morning, but he knew that if they did, the birds would be frozen solid. As much as he wanted to nestle into his bedroll and fall asleep, the work was too important. His hot breath formed clouds in the air as the sky grew dark. Barely visible through the mist was Mount Satseto, a twenty-one-thousand-foot summit that had towered over them all day, drawing increasingly closer. It was beautiful, and Ted was reminded of lines from a poem by the English author and baroness Vita Sackville-West:

> Bitter escarpments cut by knives of wind,
> Eaves of the world, the frightful lonely mountains,

> Or in Yunnan and Sikkim and Nepal
> Or Andes ranges, over all this globe
> Giant in travelled detail, dwarf on maps.

The poem spoke to Ted, and he marveled at how perfectly the words mimicked his experience. Here he was in Yunnan, sitting in the eaves of the world. Poetry did not make his knife move quicker but at least it eased his mind. Then Jack joined them, and the work began going faster. With a practiced slash of the knife, he peeled the bird skins cleanly from the bodies.

When the explorers woke the next morning, the air felt like a thousand needle pricks on their bare skin. It was so cold that Kermit's teeth began chattering before he even left his bedroll. Kermit was packing up his things when he heard yelling. "What is it?" he asked one of the guides. Everyone seemed to be running around madly, searching the barren land erratically. Finally Jack came up to him and explained.

"Mules are gone," he said, "disappeared last night, and with at least half of our supplies."

It was the only news that could make this morning worse. Already Kermit's body felt stiff and miserable. The steep climb of the day before had wrecked his muscles. His breath was ragged in his lungs from the altitude. It felt like every inhale was a gasp for air. His dirty clothes were tinged with the blood of birds, and now he did not know if he would even get breakfast.

Immediately the party divided up and began searching for the wayward mules. Usually the supplies were unpacked from the mules at night to give the animals a proper rest, but they had left the mules as they were the night before because of the rough nature of their campsite and the howling wind that came down the canyon in blasts. The guides had deemed the extra weight more insulation than burden in the cold climate. Now, of course, they regretted it.

Many of the guides were brand-new to the expedition. A stop in Li-kiang, a small but beautiful town with cobblestone streets and navigable aqueducts spanning a network of canals, had initiated a swap of guides. Some men and women were returning home, and others, mostly Tibetan women, had signed on for the adventure ahead. Ted noted their "fine features and rosy cheeks," and wrote, "If their Saturday baths came around more often, they would certainly have been both attractive and good to look at." It was a ludicrous comment, given the state of Ted's own scraggly beard and the layers of dirt that were caked in his hair. In any case, he wasn't hiring the women for their beauty. It was their skill that the Roosevelt brothers needed desperately.

The Himalayas were not a mountain range to be taken lightly. Stretching across five countries—Nepal, China, Pakistan, Bhutan, and India—the Himalayas comprised the highest peaks in the world, including, of course, Mount Everest, topping out over twenty-nine thousand feet. However, in 1929, no one could be sure which peak in the Himalayas was truly the highest. A 1924 expedition had seen British climbers George Mallory and Andrew Irvine attempt to summit Everest's peak and never return. No one knew if they had made it to the top. The only certainty was that the Himalayas had proved too much for them, as they had for so many other explorers. The summit would not officially be reached until 1953 when Tenzing Norgay and Edmund Hillary made the first documented ascent.

The Roosevelt expedition was nine hundred miles from Everest's base camp, but that did not mean their route was without danger. On the contrary, Ted and Kermit thought it possible that some of the snowy peaks they saw in the distance could be higher in elevation than Everest. No one had ever attempted to climb or survey these Himalayan peaks before, so there was no way to know. The only thing they could be sure of was that the route was snowy, cold, and steep. Only local guides familiar with the Himalayas would be capable of getting them through this treacherous section of the trail. They needed the best of the best to make it through, and so more than half of the guides were now women.

Another member had joined the expedition, a large black dog that belonged to one of the Tibetan guides. The animal was sweet and playful, and because Ted could not pronounce his Tibetan name, he called him "Bob," after the sheepdog protagonist in a favorite children's book. The new recruit lived up to his nickname in action too: he ran behind and around the mules, snapping at their hooves if they wandered off or moved too slow. Not even Bob, however, could protect the mules all night.

Ted and Kermit waited in the camp while the guides searched for the lost mules, ponies, and supplies. They didn't trust themselves out there in the ice and snow. Every direction looked the same, a blank expanse of white, and they knew that if it was up to them to navigate, they would be as lost as a mule in the mountains. At camp, however, the wait was long, and Kermit was filled with "gloomy forebodings." Finally the guides began reappearing with a few stray mules, most of the animals still missing. They took inventory and repacked their supplies. A few guides decided to stay behind to keep searching, while the rest of the party would move ahead.

When they finally hit the trail, it was already afternoon and they had only a few short hours of daylight to make as many miles as they could in the hazardous conditions. The wind blasted through the canyons with so much force that Ted felt he was going to be carried off by it. He sputtered as the wind scattered a spray of ice and dirt across his face, then looked over at the guides. They seemed to hardly notice it. "Tibetans," he wrote, "seem impervious to the wind." What Ted could not appreciate was that the gusts were a fundamental part of Tibetan lives and, as in neighboring China, their presence as constant as the earth beneath their boots.

Just seventeen years earlier, Tibet had expelled the Chinese from their kingdom, in a burst of violence that followed centuries of conflict and occupation. That year, 1912, the Chinese had officially surrendered, making way for the Dalai Lama to return to the Kingdom of Tibet after spending six of his last eight years in exile. "I, too, returned safely to my rightful and sacred country," he proclaimed in 1913, "and I am now in the course of driving out the remnants of Chinese troops from DoKham in Eastern Tibet. Now, the Chinese intention of colonizing Tibet under the patron-priest relationship has faded like a rainbow in the sky."

The rainbow, however, was persisting. Border disputes continued, and in 1914, Tibet advanced to the east, defeating the Chinese in a series of skirmishes and claiming more land for their own. The British sent their support to the nation, assuring Tibet that they supported the country's autonomy, but it wasn't enough. By the late 1920s, Chiang Kai-shek's new nationalist government was extending its reach. Tibet was in its sights, viewed as an outdated relic, whose leaders were "tyrannical" and whose citizens deserved modern Chinese governing.

Straddling a border visible only on their maps, and with lines that differed depending on the cartographer, the Roosevelts found it impossible to determine which lands were Tibetan and which were Chinese. The mountains and valleys looked the same, but Tibet was undoubtedly distinct from its neighbors. Its people's customs were different, their food rich with yak butter, and their ability in the mountains unparalleled. And none of the women had bound feet; the bone-breaking custom had been left behind on the border. Now that the expedition had entered what Ted called the "region of perpetual snows," the Roosevelt brothers knew how fortunate they were to have them as guides.

Not that they felt lucky in the moment. The afternoon presented a brutal climb in slippery conditions, with no hope of shelter. As the evening hours approached, the guides found a hollow partially protected from the wind. The men and women set up their tents quickly, anxious to find relief from the stinging gusts that blew down from the peaks around them. A creek gurgled noisily next to their campsite, runoff from the crystalline slopes of glaciers above them. The water shone golden as it cut through the snow, reflecting the dying light of day.

Kermit had a headache and lay still in the tent, but Ted and Jack remained intent on their specimen collection despite the conditions. They set out a line of rodent traps, hopeful that they might uncover interesting small animal species in such a remote region.

It was an uneasy night's sleep for the expedition. The wind was so loud and powerful that it was hard to rest at all. It came "thundering through the tree-tops" first, a sound so loud it would rouse the Roosevelts, especially because they knew what followed. The tumult in the trees was a harbinger of the blasts that would soon try to devour their tent. The canvas of the tent would be pulled down, coming within inches of their faces, the force occasionally so strong that it would collapse the entire structure on top of them.

Morning could not come fast enough. Finally the first rays of sunlight came shining down, and the explorers hesitantly left their tents. Ted and Kermit gasped when they saw the world outside. Frost covered everything, even their own beards and hair, and when the sun's rays hit the ice crystals, the world glittered irresistibly. The mountainside had been transformed into a realm of sparkly diamonds. Unfortunately, there was ugly work to do in this beautiful landscape. Kermit, Jack, and Ted's numb fingers were stiff as they examined the two species of mice their traps had caught the night before. They wiggled their fingers to warm their blood, and then set to work.

As the team was preparing specimens, the guides from yesterday's misadventure returned to camp. They had trekked across the snow, exploring as far as they could the day before, but finally had to accept failure. The mules had disappeared. The new, falling snow had covered their tracks, and now the animals were lost somewhere in the wilderness. It had been an exhausting search that extended into the freezing evening hours, and it had ended in nothing. There was no time to rest. The expedition had to keep moving, especially since their supplies were low. Most of the food they had bought in Likiang was now frozen somewhere on the back of a lost mule. Worst of all, there was no village ahead to replenish their supplies. They had two weeks of tough, arduous climbing ahead and no food in sight.

As they started up the trail, a low thudding pain radiated through Kermit's head. *What else,* he wondered, *could go wrong?* The universe was about to answer.

CHAPTER 5

HOUSE OF THE PRINCE

It was altitude sickness; everyone was sure of it. The signs had started slowly, headache, fatigue, and an odd, muddled thinking, as if one were making decisions submerged in feet of water. These symptoms, Kermit knew, could be anything. They could indicate a viral infection or even dehydration. But the nondescript illness became different in the middle of the night. That's when Ted and Kermit woke up choking. "You wake every few moments," Kermit wrote, "struggling for breath, and feel as if you had been long underwater." Their bodies heaved in desperation for one simple gas, oxygen, plentiful in the valleys thousands of feet below them but severely lacking in the mountains. It was a terrifying way to wake, as if you were dying and your futile, hacking breaths served as a final alarm clock.

It was the last day of January 1929, and the past week had proved disastrous. They had lost more than half their mules and supplies, in two separate incidents where the animals wandered off into the mountains in the middle of the night. Then, while sitting around the campfire, Suydam had stirred the embers, sending a shower of sparks toward Kermit. Thankfully there were no physical injuries, but the web of ragged holes in his jacket left Kermit cold and miserable. It was his only jacket, and the timing could not have been worse. They were reaching a high altitude, 15,000 feet, and the air was growing increasingly wintry and thin. "What might be an easy climb at 10,000 feet," Kermit wrote, "at 17,000 sets the heart beating like a trip-hammer and the lungs gasping for air."

As the hail caught in their hair, there was little time to consider altitude sickness. Even though their bodies were crying out for them to stop and rest, the explorers rose early, before dawn. They were climbing a steep and tortuous ridgeline known as Tiger Leaping Gorge. The area was sometimes referred to as China's Grand Canyon because of its immense chasms, stunning thousand-foot drops off sheer rock cliffs, and spectacular views. The explorers reached a frosty meadow by midday and the guides noticed the natural protection from the wind that the land offered them. It was a perfect spot to camp. Below them were thick pine forests buried under a blanket of deep snow. In every direction they were surrounded by one of the most biologically rich areas on earth. The beauty was overwhelming, and in the moment Ted, Kermit, and Suydam knew exactly what they wanted to do.

The three men took off separately, each carrying a shotgun and a mental list of the mammals that the Field Museum in Chicago wanted for their collection. It was a wish list of near-mythical creatures, including one whose whereabouts, habitat, and behavior were still unknown. Truthfully, any mammal would be welcome. This expedition was expensive, and everyone knew that the likelihood of them snagging the big prize, the panda, was small.

Kermit made his way up through the snowy forests, to the edge of a ridgeline. He came across the fresh tracks of a sambar, a large native deer, and began to track it. Up Kermit climbed for more than an hour until his legs burned and his breath became tattered in the thin mountain air. He could find no deer. The trees thinned out and soon he was scrambling up a rocky hillside with a vista across the mountains. It was there that Kermit saw an animal silhouetted against the rocks. It wasn't a sambar, but instead a serow.

Locals had talked about this creature, calling it a "mythical wild goat." It was an animal so shy that none of the villagers had ever seen one. Little was known about it beyond the fact that it wandered alone in the Himalayas. Kermit, surprisingly, knew more about the animal than the people who lived among it. He had at least heard it described by Roy Chapman Andrews, the famous naturalist, who had shot a specimen of the serow for the American Museum of Natural History in New York in 1916. Even better were the pictures that Roy's wife, Yvette Borup Andrews, had taken during their joint expedition, a treasure trove of landscapes, botany, people, and, of course, highly sought-after mammals.

Yvette was prodigious in her fieldwork, able to hike great distances, and photographed her subjects, both wildlife and cultural, with sensitivity, even pursuing innovation. Impressively, she developed her own images in the field using a portable rubber darkroom. Despite this, her husband was dismissive of her work following their 1931 divorce. "Physically and intellectually," he said to the press, "women may be the equals of men for the work of exploration, but temperamentally they are not. They do not stand up under the little daily annoyances that loom large to them in the somewhat trying work involved on an expedition. The trivialities which men can ignore completely disturb them and prevent them from settling down to hard and conscientious work."

It was a preposterous statement, likely driven by Chapman's jealousy and bruised ego after his wife left him for another man. Ted and Kermit's philosophy was starkly different, influenced by their father, who was at the forefront of advocating for women's rights. "Much can be done by law towards putting women on a footing of complete and entire equal rights with man—including the right to vote, the right to hold and use property, and the right to enter any profession she desires on the same terms as a man," he wrote in 1913. "Women should have free access to every field of labor which they care to enter, and when their work is as valuable as that of a man it should be paid as highly." While no English or American women accompanied the expedition, the Roosevelts did pay both their male and female guides equally.

Even with both men and women guiding them, the Roosevelt expedition was struggling with its own set of "daily annoyances" and "trivialities," so finding the serow felt like a triumph to Kermit. This animal was slightly different than Andrews had described, but he wasn't at all in doubt. The serow's coat was dark brown and its horns long and curved. Around its neck was a strange tuft of hair, almost like a scarf looped around. It was 150 yards away and Kermit drew up his shotgun. He held the gun steady and fired, but missed.

Most mammals, except for humans, use smell as an essential part of their social behavior. The serow lives in isolation using sniffing and scent marking. Each animal roams its own piece of the mountains, careful not to invade another's space unless as part of a mating ritual. The serow prefers to be far from its own kind and seeks the solitude of the mountains. At the smell of any other animal, it runs.

Few social behaviors are as important as territoriality in mammals. It is intimately tied to violence. The need for an animal to defend its territory against others of its kind transcends species and geography. It closely shapes how animals interact with their environment, from the elusive wanderings of the serow in the Himalayas, to the aggressive pursuit of prey by a wolf pack in the Arctic tundra, to the ostentatious barks of male elephant seals on sandy beaches. Humans, of course, are no exception.

Kermit was far enough away, with the wind in his face, carrying his scent away from the serow's nasal sensory cells. Quickly he reloaded and fired again. This time he hit the animal. The serow staggered in the snow. He reloaded. His third bullet killed it. Kermit felt a deep satisfaction. They had bagged a large mammal for the museum.

The expedition was headed next for the Kingdom of Muli, an independent territory ruled by a royal Tibetan family. In the West, Muli had been dubbed "the Lost Kingdom" in *National Geographic* articles, due to its secluded location and the exceedingly dangerous, rugged trail that led to it. The area held little importance, economically or strategically, and so was mostly left to its own governance, at least for the time being.

The eastern ridge of the Himalayas created a physical barrier for the ancient cultures harbored in its remote plateaus. There were the Tsang (Tibetans), the Han (Chinese), the Khoshut (Mongolians), and the Hui (Muslims). But there was also a wealth of other minorities, including the matriarchal societies of the Mosuo and the Nashi, as well as the Yi people, whom the Han called "Lolos" and who were sometimes held as slaves by others in the region. Altogether, the diversity of wildlife was matched by a vibrant assortment of cultures and people. Some believe it is this region, a mix of isolated peoples in lush valleys, that inspired the mysterious paradise of Shangri-La imagined by English author James Hilton in his 1933 novel *Lost Horizon*.

"One of the least-known spots in the world is this independent lama kingdom of Muli," botanist Joseph Rock wrote in an article for *National Geographic* in 1924. "Almost nothing has been written about this kingdom and its people who are known to the Chinese as Hsifan, or western barbarians." The Roosevelts had heard so many varying accounts of the Kingdom of Muli that it was difficult to know what was real and what was hyperbole. The Christian missionaries told them that Muli was anti-Chinese and therefore "unfriendly" and "savage." Joseph Rock, whose articles on the "land of the yellow lama" described the spiritual leaders in Tibetan Buddhism, gave Ted the most practical advice: "Bring gifts."

The sky was dark by the time the serow was carried back to camp and Kermit was too tired to prepare the specimen properly. He knew it was a mistake, one that he would regret later, but overwhelming exhaustion flooded his body, and he could do nothing but crawl into his bedroll and plunge into a deep, dreamless sleep. He was too tired to eat dinner, not that there was much food left in their supplies. Bags of dried green peas and rice were all that either the explorers or the mules had to eat, and both humans and beasts were sick of both options.

The next morning Kermit stirred late, after the others had already woken, and then remembered the serow. He dragged himself out of his warm, cozy bedding and trudged over to the dead animal that he had left lying in the snow the night before. When he prodded the serow with his fingers he realized that his fears had come true. The animal was frozen solid, "stiffer than any Argentine steer in cold storage." In his burnt jacket, the hail prickling through the holes in the fabric, he felt like he could easily be next. Ted and Suydam were off hunting in the surrounding hills, so it was just Kermit stuck with the misery of skinning a frozen serow, a task that was much more difficult than attending to the specimen when it was fresh, in the same way that it's harder to cut a frozen chicken breast than a fresh piece of meat. All the elation after the hunt yesterday had drained out of Kermit as he wielded his knife with a cramping, stiff grip.

It was afternoon before the explorers packed up and hit the trail. They were each dreaming of the same luxuries: warm food and solid walls. They were so far away, more than fifty rugged miles, from the nearest town. But then a mirage appeared: a stone building whose walls seemed to rise from the rock around them. The men and women blinked at the structure, unsure of what they were seeing, and then hiked as fast as they dared, desperately hoping that it was real.

The trail teased them. It twisted this way and that, keeping them a maddening distance from the stone house, but then they began to hear the beat of drums. Kermit said that it sounded like "thunder," echoing off the mountains, and they found themselves matching their steps to the noise as the trail stretched on and on. The pounding of Tibetan drums drove them, giving them hope that people, warm beds, and food lay ahead. It grew louder as they approached.

Finally they reached the stone house. They stared at the exterior with wild eyes, disbelieving that it was real. The solid building could no longer be denied. It was three stories high with an angled roof, the tiles set in ridges, one rising impressively above the last. Strings of Tibetan prayer flags fluttered in the wind, sending their blessings across the countryside. It stood alone, far from any town or any visible residents, of whom the only evidence was the loud, deep pitch of the drums, whose rhythm was now integrated into their every movement—a rhythm that, in fact, kept time for the lamas, who used the sound to assist with their meditation.

As Ted approached the front door, he noticed a monkey chained near its entrance. The animal, miserable in its captivity, lunged for Ted, "sinking his sharp teeth" into his leg. It was an unhappy beginning, but once the doors were thrown open, the expedition realized how lucky they were. They had stumbled upon a lamasery, and the elation they felt as they made their way into the hall had nothing to do with the sunny, golden robes of the monks who resided within.

At the age of seven, children perceived to show "a sacred sign of the buddha" would enter the lamasery. Those signs—a pronounced ridge at the base of the thumb and loose skin between the fingers—were a ticket to a life of celibacy and chastity in service to the community. Lamas—both men and women—were known for their skill in medicine, often serving as the only doctors in the areas where they lived. While some lamaseries were single-sex, the one the expedition had found was well supplied and housed all genders.

The Roosevelts, Suydam, and Jack were led through hallways covered in detailed murals and into a spacious courtyard where a pagoda sat, housing a statue of Tsongkhapa, the founder of the yellow sect of Tibetan Buddhism. All around them stood a group of lamas dressed in yellow robes, warmly greeting the newcomers. They were immediately welcomed to stay in the lamasery, and, much to the explorers' delight, offered cups of tea.

The Tibetan tea was flavored with yak butter and salt, and Ted drank his greedily, enjoying every sip. He thought it more like soup than tea, but the name was unimportant when hot liquid reached his belly and the warmth spread, encompassing his entire body. He sighed in pleasure, but the best was still to come. When the lamas offered them food, clean bedding, and supplies for the trail ahead, it was as if their wildest dreams had come true.

One of the men noticed the burnt holes in Kermit's coat and took him aside. He offered him the yellow fabric that made up the robes of the lamas, and together the two patched the coat in squares of bright yellow. "I'm part lama now," Kermit said with a laugh as he admired their handiwork. The yellow robes symbolized the way the golden leaves fall to the ground in autumn. Wearing yellow was a reminder to never let go or give up. He didn't know it yet, but it was a lesson that Kermit would need in the months ahead.

A few days in the lamasery provided all the rest and supplies they could wish for and renewed the explorers as they returned to the trail. They needed it, as the landscape was not going to become more forgiving. A couple days were spent hiking in rugged country, and on the third day they climbed a steep pass. Reaching the top, they saw the Kingdom of Muli nestled in the valley, mountains rising on the other side, like a bird's nest resting in the branches of a tree. Even without their mules and supplies, and on the verge of dangerous altitude sickness, they had made it.

The winds were picking up, fierce and cold in their faces as they marched, yet could not negate the beauty of the landscape below. Massive black firs and hemlocks framed their view of one of the oldest stands of trees in the Tibetan Plateau, their heights reaching to over a hundred feet, with trunks that spanned five feet. Although it was early February, the alpine meadows were dotted with flowers. Delicate white anemones, called windflowers, lined the trail, their petals wafting in the breeze.

The trail zigzagged haphazardly until finally the city of Muli appeared below, its stark white buildings contrasting with the deep-green evergreens around it. A wall of whitewashed stones surrounded the city. They were entering an enigmatic land, cut off from the rest of the world by steep mountains and craggy trails, and known for its isolation and its tyrannical king. As Kermit peered down at the white city, he wondered what they would find.

It was night before the group reached Muli. All was quiet with not a soul to be seen. The February air was frosty as the travelers made their way through the empty streets and Kermit shivered miserably. Even with his yellow-patched coat, he couldn't seem to get warm. The white temples of Muli glowed eerily "ghost-like" in the dark. Every building in town had been painted a stark white and the moonlight cast a bluish haze across the scene. Kermit was not a man easily spooked, but the setting disturbed him. He longed to be inside by a fire.

HOUSE OF THE PRINCE

City of Muli. Photograph by Suydam Cutting, 1928.

The next morning, under a deep-blue sky, the explorers made their way to the House of the Prince. They had sent the guides first, armed with modest gifts, so that their arrival would be expected. From the outside it was obvious that this dwelling was different from any other in the city. It was three stories, wide at the base and narrow at the peak, and with windows shaped like triangles. A wall of rough tree trunks surrounded the structure. Because the king of Muli was away traveling, the party had chosen to stop at the home of the king's brother. As they approached the house, a swarm of men appeared, blocking the stone-paved road. With a swift movement, the dozen men bowed deeply and then stuck their tongues out at the Roosevelt expedition. Ted smiled back; it was a typical Tibetan greeting.

As they entered the house, the party was struck by its beauty. It was by far the most elegant home they'd seen in months. Ted's eye immediately caught on a large painting of the Buddhist Wheel of Life that hung on the wall. The image was familiar—they had seen it represented in artwork before and hung in homes and temples. In bright, beautiful colors, the image shows the six realms of existence, or states of mind. The cycle of birth and reincarnation is depicted, along with the path to nirvana. The Wheel of Life is painted on the outside of nearly every Tibetan Buddhist temple and is meant to be accessible. It explains Buddhism in a way that anyone can understand, even those who are illiterate or, like the Roosevelts, oblivious.

House of the Prince. Photograph by Suydam Cutting, 1929.

In the middle of the wheel, ringed by the six realms into which a soul can be reborn, are the three causes of suffering: ignorance, greed, and hatred, represented by a pig, a bird, and a snake. On the outer rim are shown the twelve states of dependent origination, a detailed look at karma's role in life and how every moment of existence is linked to our past, current, and future lifetimes. At the bottom, holding the wheel, is a figure representing impermanence. At the top is the Buddha, pointing to a full moon, representing liberation from the cycle, or enlightenment.

The wheel depicts how humans become trapped in a never-ending cycle of suffering and explains how to stop the process. Buddhists believe in cause and effect. Every selfless action brings meaning to life. The steps show that altruistic purpose is not just about helping others but is essential to one's own happiness. It is through selfless acts that humans can be released from the wheel.

Liberation from the cycle, or nirvana, represented by the moon, is not a well-trod path marked with road signs. It's an expedition that requires the underbrush to be cleared, the trail to be discovered and then followed, even through all its twists and turns and despite the fearsome animals to be overcome along the way. Everyone's path is different, but it leads essentially to the same place.

As Ted and Kermit admired the painting, they were absorbing lessons that they weren't fully ready to contemplate. Instead, they were overcome by the beauty of the piece. It was exquisitely detailed, and Ted stared at it greedily, desperately wishing he could buy it. He was used to purchasing things he liked, and in general getting the material possessions he wanted. He loved shopping, but this was no marketplace. As assured as the eldest son was, he was not so coarse that he would tramp into a person's home and inquire about buying their belongings. He worried that even asking about the beautiful painting might be offensive.

Their host appeared. Through their translators, he introduced himself as Chang Ta Li, guardian of the eastern border. He was a tall, strong man with hair cropped short and his head deeply scarred. A small, round red cap hid some of the mutilations, and he wore long, flowing red robes. Immediately Ted presented the gifts they had so carefully brought from the West: a turquoise ring, an emerald-green hat, a flashlight that was powered by a hand crank, and a .410-bore shotgun, ideal for small-game hunting, complete with ammunition.

Chang Ta Li seemed more interested in photographs. Ted pulled out the family pictures he carried with him, and his host closely examined a photograph taken of three generations of Theodore Roosevelts: the president, Ted, and Ted's three-year-old son. The guardian was charmed by the image of Ted's young son. He stared at the photo with a wide smile, pointing to the boy. He then introduced his own family: his wife and chubby, boisterous three-year-old son, who would one day become king of Muli.

The king of Muli was a lama ruler and so bound to a life of chastity. He could not marry or bear children of his own. However, because the position was tied to the royal family, this meant that the eldest son of the king's brother was next in line for the crown. Chang Ta Li, although not a king himself, was raising one. The house that Ted admired did not truly belong to Chang Ta Li. It was called the House of the Prince because it belonged to the sweet toddler who was currently running noisily around the room.

Chang Ta Li showed the various photographs of the Roosevelts to his wife and son, remarking on the large family. Then he pointed to the elder statesman, asking Ted pointedly, "Who is President Roosevelt?"

CHAPTER 6

SOUTH OF THE CLOUDS

The Roosevelt brothers had always, since their earliest days, stood in the long shadow cast by their father. It was part of their identity, an essential component that influenced their every action and their demeanor toward others. Now this core part of their being had been stripped away. Ted and Kermit were in a part of the world where American politicians meant nothing, a place where they would be judged on their own actions instead of their father's.

It was a strange, weightless feeling to realize that all their family's wealth, importance, and influence had dissipated into nothing, vanquished by geography, a gate made of sixteen-thousand-foot-tall mountain peaks, and a trail so rugged that few residing outside the region's borders had ever walked along it. "It has always been remote and inaccessible," Peter Matthiessen wrote of the area they now inhabited, "more so than any land on earth."

Chang Ta Li did not know who President Roosevelt was and had little use for America. With the Roosevelt brothers, he wasn't interested in talking about powerful men and their accomplishments, as had dominated the conversation for Ted and Kermit over so many continents.

From right to left, Theodore Roosevelt Jr., Jack Young, Chang Ta Li with his son, and Kermit Roosevelt. Photograph by Suydam Cutting, 1929.

Instead, he wanted to learn about them. The shards of the Roosevelts' lives came crashing down around them. Ted was no longer the failed politician, accused of corruption by strangers, colleagues, and merciless family members grasping for their own political power. Kermit shed his inability to hold on to success in his business and his shaky marriage, and the insecurities that stemmed from the harsh words he had sometimes heard from his father, who'd once accused him of being a "weakling." In Muli, the explorers were reborn, able to taste life without the burden of their past.

It was February 10, 1929, and the explorers found themselves at an auspicious location to celebrate the Tibetan New Year, or Losar. The entire kingdom was bursting with celebration, and red decorations hung around the grand house. The guardian called the Roosevelts, Suydam, and Jack into a room of the house and invited them to sit on a divan covered in bright-colored fabric. His wife and son were there, and the family scene was easy and relaxed. He gave each of the men four oranges and then began to pour what looked like water into small bowls on a low table. Ted looked at the drink suspiciously. When it was time to drink, he and Kermit purposely chose the two smaller bowls on the table. They raised their cups in celebration of the new year and then drank. A burning sensation ran down Suydam's throat and it took all his grace not to sputter the moonshine across the room. "The liquid that we tasted was so strong," Ted wrote, "that it nearly burned our lips."

After their surprise morning beverage, the group moved outside, where Ted and Kermit distributed gifts to their guides and the rest of their team to mark the occasion. It was a day of rest and celebration, and the explorers roamed the village, taking in the cluster of small houses and the lamasery. Unlike the one they had stayed in a few days prior, this building was massive, housing some seven hundred lamas, and richly decorated. A golden statue of Maitreya, the future Buddha, towered over them. It was fifty feet tall and so beautifully sculpted that it took one's breath away just to stand in its presence. The glory of the statue's golden skin stood in sharp contrast to their own. No matter where the men went, they drew odd glances. It wasn't just that they were white men and outsiders—their long beards and grimy hair and skin made them a spectacle.

Lamas in Muli. Photograph by Suydam Cutting, 1929.

Thanks to the New Year celebrations, a happy mood pervaded the village, contrasting with the poverty of the region. While the royal family enjoyed luxuries and ample food, most people in Muli endured long days, hand-sowing wheat and barley in rocky soil, herding yaks in the hills, and tending to watermills on the creek. To the Roosevelts, blinded to the austere realities, the kingdom reminded them of a fairy tale. Rows of soldiers in crimson coats lined the path to the House of the Prince. They even carried muskets, weapons from another era.

Days in the House of the Prince passed too quickly. Ted and Kermit wished they could spend weeks in its halls, with its delicious meals and good company. The guardian of the eastern border urged them again and again to stay longer. Yet as comfortable as they were, particularly after the struggle of the last few weeks, the trail called to them. They were three weeks away from what some locals were calling "panda country," and they knew they couldn't linger, no matter how restful they found Muli. Ted and Kermit were "really sorry to leave," but ahead lay all their ambitions.

As the party took leave of the House of the Prince, Chang Ta Li approached Ted and Kermit with presents. Among the gifts he gave Ted was the one painting he hadn't dared to ask for: the Wheel of Life. He was overcome with gratitude. Ted couldn't understand how Li had known how much he wanted it; he had thought his behavior was subtle, and only Kermit knew how greatly he desired the painting. Its presence felt like fate, as if the painting had chosen him. As they left the house and marched down the long street, Ted turned back. In his bright scarlet clothing, Chang Ta Li had become a tiny red dot on the horizon. The guardian was still watching them.

Ted knew that they needed someone to be looking out for them. As difficult as the last section of the trail had been, this next climb would be far more dangerous. They were exploring a section of the Himalayas that was little traveled due to its punishing route up a seventeen-thousand-foot peak. All their struggles in the last mountain range would be magnified here, especially as the trail tended to hide from its followers. The path was so little trod that it became easy to lose, diverging into yak tracks that slithered this way and that before ending abruptly in the middle of an icy meadow, the yak having decided to turn around and go home, leaving the traveler hopelessly lost.

They weren't in the snow yet. As the explorers climbed out of Muli, they entered a vast clearing. They noticed that some of the pine trees had been scored to extract the sap. It was a common ingredient in traditional medicines. Because this wilderness had felt little impact from humans, the diversity of plants was exceptionally high. There were no clear-cuts this far out, for no one could haul logs down the precipitous trail. No large settlements of houses had blossomed out from the city center, so large tracts of land had not needed to be cleared to support a growing population. The wilderness had run free, and an overabundance of new and rare species could be found wherever they looked.

The group passed a forest of rhododendrons. There were so many different species contained in this one section of the world that they had not all been counted. However, thanks to the botanist Joseph Rock, a few prized samples had been sent to the United States, and some of the same extraordinarily rare flowers that they passed on this hillside could also be found in Harvard's Arnold Arboretum in Boston, where they flowered as brilliantly for city dwellers seven thousand miles away as they did for the Tibetan blackbird on a remote hillside.

The group was headed to Tatsienlu, a small town some two hundred miles to the northeast that the guides translated to Forge of Arrows. Although the explorers still had yet to meet anyone who had seen the panda, it was this part of the world that most scientific expeditions were bound for. This was where in 1919 the missionary Joseph Milner had bought a panda skin, donating the item to the American Museum of Natural History and setting off a fervor among scientists and explorers that the black-and-white bear was real. Despite a decade of failed explorations since then, it was here that the Roosevelts felt their best hopes lay. "Slight as our chance might be," Ted wrote, "we felt that we must make every effort to get to a country where panda existed."

Ted, Kermit, Suydam, and Jack had stayed behind the caravan, savoring a last good-bye with Chang Ta Li, and this meant that they now had to hurry up the trail to find their guides and mules. A lama was generous enough to accompany them, helping them find their way along the narrow path that led through meadows and forests into the mountains. The task proved difficult, even for him, and by afternoon they were frantically lost. Tracing the path was nearly impossible and they found themselves at the end of a yak path to nowhere and obliged to turn around.

At nightfall, as their anxiety reached a peak, they saw the smoke of a campfire ahead in the valley. They marched faster in the dark, turning their ankles on the path, but happy to have a destination. The group was camped in "a pretty little hollow" and the mules had been set free to graze, as they always were at night, with their packs lined up in circles to create a windbreak for the fires. The campfires were blazing tonight, thanks to a combination of plenty of wood and little wind, and the explorers sat around, drinking buttered tea and listening to the singing of the guides. Ted was mildly annoyed by the "shrill" singing in "monotonous cadences," but everyone else seemed to appreciate the song. Suydam began humming along, even though he knew neither the melody nor the words. Everyone knew this was their last camp before they crossed into the snowy peaks, and they were determined to enjoy it while it lasted.

The next day the explorers would enter a "harsh wasteland," where icy, lashing winds cut at their exposed skin. The snow piled high in the valleys and the guides struggled to "break" the trail, re-creating the path that had disappeared in a fresh coat of white, ahead of the rest of the group. The mules, even those newly acquired from Muli, struggled to climb the large snowdrifts that had blown across their path. The caravan's pace was slow, and as evening approached, they could do little but pick the closest patch of white in which to make a campsite. They had climbed to a pass of fifteen thousand feet and immediately everyone could feel how thin the air was. There were no trees up here, merely a view of two snowcapped peaks.

Dawn after a cold, windy night in camp. Photograph by Suydam Cutting, 1929.

The cold penetrated their bodies so deeply that even sitting by a fire could not thoroughly warm them. "One side would be cooked and the other frozen," wrote Ted. "Finally, we crawled into our bedding rolls and lay there shivering. Not so the Tibetans . . . cold meant nothing to them." The Roosevelts no longer bothered to change their clothes at night, dreading the feeling of cold air on their skin, instead piling on the blankets, trying to create a snuggly warm nest. They lay cocooned in as much warmth as the expedition could provide, and even that was not enough.

The guides often had no blankets at all. They slept in their clothes, with no sleeping bags or bedding to make the hard ground more comfortable or warm them against the chilly air. While the men and women leading the group slept peacefully, Ted and Kermit woke in the middle of the night with the same telltale clutch of the chest as before. They were choking, a feeling like drowning, even though they were miles from any body of water. The air was too oxygen-deprived for them to breathe, and unlike their guides, their bodies hadn't adapted to the altitude.

"We were almost exhausted when the top of the pass was reached," wrote Jack. "Deep snow lay all around us and the wind increased in violence. Our mules arrived half dead; and there was no shelter for the night. With numb fingers we pitched our tents on top of ice. Our thermometer registered several degrees below zero. We melted snow for tea and after supper we crawled into our Fiala sleeping bags."

As much as the humans suffered, the mules had it far worse, so perhaps it should have come as no surprise when one morning, soon after leaving Muli, a third of them simply disappeared. "Personally I did not blame them," wrote Kermit. "Lying in blankets or huddling by the fire was bad enough. Standing up sheltered in the wind and driving snow must have been unbearable, even for a salted mountain mule."

It was the worst news. They had only just replenished their supplies and mules since losing half of them in the mountains before arriving at Muli. Now they were in the same precarious situation with half of their food stores vanished. The situation was dire. The guides went out searching across the mountains, but the winds blew them back to where they started. The wind was so strong that one could lean completely on it, as if the air were made of stone. Still, they had to keep trying. A search party of men and women set out again to look for the missing mules while the rest of the explorers continued on the trail, even though they knew their progress might be for nothing if the animals weren't found. The area they were traveling to was so remote that they simply could not hike onward without their food. There weren't enough villages along the way to supply them.

Besides the supplies on their backs, the mules were needed for another purpose. Their heavy hooves helped break the trail for the guides, and without them the path was nearly impossible to follow. To prepare the trail for Ted, Kermit, Suydam, and Jack, the guides had to hike far ahead of the men, pounding down the snow with their fists. It was exhausting work, hard on the back and the legs, and the driving snow stung their faces "like birdshot."

The remaining mules on the expedition were weak and miserable.

They fell into snowdrifts or slipped on patches of ice, in one instance requiring the party to physically haul an animal back to the trail. To Ted it seemed the mule was "deliberately trying to kill itself by leaving the trail near a precipice." It took all their strength to lift the packs off the animals and guide them back to the trail being made by the pounding of human hands. Bob the dog ran in circles around the mules, trying his best to keep them in line.

To Ted and Kermit, the land appeared barren, devoid of all life, either plant or animal, but Suydam making his way on the snowy trail had just spotted something. He moved in closer to the face of a cliff, where a strange-looking plant was nestled in among the rocks. It looked like an alien, covered in green and purple and resembling a hairy starfish. Despite the scientist's delays, Suydam missed Herbert, and he worried about him too. They all wondered how the botanist would fare in these mountains without the team of guides to help him. Knowing how important plant collection in this part of the world was to Herbert, Suydam scooped up the plant and decided to add it to their collection.

It would turn out to be a Himalayan snowball plant, *Saussurea laniceps,* an unusual species with a green, spongelike body covered in purple rosettes. The plant is found only at high altitudes and is called a "woolly plant" because it's covered in a thick layer of trichomes, tiny structures that some scientists believe act like hairs on a plant to keep it warm in the deep snow, although their exact function is a mystery. The plant is prized in China, however, not for its hairy leaves but for its role in medicine.

When properly harvested and prepared, the Himalayan snowball plant is an effective treatment for arthritis and stomachache. This isn't surprising, as it is part of the Asteraceae family. Members of this family of flowering plants play a critical role in traditional and modern medicine, with applications in heart and liver disease, cancer treatment, epilepsy, hypertension, and antimicrobials, among others. The vast majority of the world, more than 80 percent, is dependent on forms of traditional medicine.

A close relationship exists between the diversity of plants and the evolution of medicine. The region the explorers were wandering in was bursting with unusual species. All around them lay the potential to treat disease and ease human suffering, if only they could recognize the rare species and collect them carefully. The Himalayan snowball plant was easy to spot; it looked so different from any other plant the group had ever seen. But many others were concealed all around them. The area they were hiking through was special, a sliver of the world virtually untouched by human activity and containing a wealth of plants more beneficial to humans than any animal the explorers could hunt.

Surrounded by the snowy Himalayas, it was hard for the explorers to look past their suffering. As dark fell, the group found a campsite at a slightly lower level in the trail where the snows were not quite so deep. They were back in a rhododendron forest, but this one had been killed by a strange blight, many miles from human hands. "Ash white gnarled trunks and limbs," wrote Kermit, "were moonwashed and goblin-like against the background of snow." The dead forest was creepy, but at least the plants provided some kind of windbreak against the howling gusts that moved around them.

Ted, Kermit, Suydam, and Jack were exhausted from the brutal day on the trail, but it wasn't time to curl up in their bedrolls yet. Not everyone had made it safely to camp. There was no sign of the contingent of guides who had gone out early that morning to look for the lost mules and supplies, sharply aware that their survival in the mountains rested on their ability to recover their food. Hours passed in darkness, until it was ten o'clock at night, and still there was no noise but the howling wind. Everyone in the camp was nervous for the men and women who were part of the search party. It was incredibly cold; temperatures had plummeted to five degrees Fahrenheit, and even next to the fire the group shivered.

To take their minds off their troubles, the Roosevelt brothers turned to their books while they waited. Unlike the other supplies, they always carried their books with them so that they could read whenever they had a free moment on the trail. For now, it was the only way to stay awake after the exertions of the day. Kermit curled up by the fire with his copy of *Pride and Prejudice* and let the comfort of well-trod words wash over him. "It is a truth universally acknowledged," the book famously begins, "that a single man in possession of a good fortune, must be in want of a wife." As he turned the pages, the dark, cold night transformed into the well-lit drawing rooms of nineteenth-century England. Instead of the howling wind, he heard the witty banter of Elizabeth Bennet.

The Roosevelts had always loved books, especially on their expeditions. The stories offered relief in dark moments, when nothing was going right and success was not guaranteed. The ability to disappear into a different world, if only for a few moments, strengthened the mind during trying times. "The lack of power to take joy in outdoor nature," wrote Ted and Kermit's father, "is as real a misfortune as the lack of power to take joy in books." Like him, Ted and Kermit could not imagine entering the wilderness without novels.

Ted and Kermit kept reading, until they heard an odd sound. They looked up from their books and realized that the snow had stopped falling and the air was clear. Then Ted heard the noise again. It came from beyond the dead rhododendron limbs that surrounded them. It was singing. The group waited by the fire anxiously as the sound grew louder. And then, moonlight washing over them, the round faces of the guides emerged in full song.

Throwing aside Jane Austen, Kermit howled in delight. The campsite was suddenly the scene of raucous celebration. Every guide was yelling out joyously as they ran to welcome the latecomers. Best of all, they had found the mules and were leading them to camp. Suydam was particularly excited to see his favorite pony: "It was a ridiculous little animal, piebald, with a mane as stiff as a scrubbing brush, and a tail that nearly trailed on the ground."

While Suydam smiled at his pony, Ted was thrilled to see one of the guides, a man named Luzon, back from the rescue mission. He was "a fine figure of a man," Ted wrote, "the grace of his proportions was unconcealed. He was never tired. . . . In the evening he superintended every detail of the care of his animals. In the morning he was first up and about. His face was battered and twisted by weather until it looked like a blackthorn. He was always happy, and I rarely saw him without a smile on his face."

The Roosevelt brothers felt a deep obligation to the men and women who had risked their lives trekking through the snowy mountains in the dark and cold. Their heroic actions had saved the entire expedition. Obscured under the Roosevelts' gratitude, like a mountain buried under snow, lay old prejudices. "Practically all natives steal," wrote Ted, just a day after the hard-won return of their supplies by their "native" guides. Nothing had been stolen; Ted was merely scribbling his thoughts, but it seemed that no sacrifice the guides could make was enough to shake his mind free of bigotry.

Trails in the Himalayas follow the ridgelines of the mountains. They sweep up sharply, rising in altitude rapidly, and then descend again, dropping to the valley below before they begin their upward climb once more. The trail was hard to follow in the snow, and the guides had to search carefully for cairns, small piles of rocks that others had left behind to mark the way.

The explorers followed the trail down into a spruce forest. There had been no logging anywhere near here, and the majestic trees rose all around them. The canopy above them acted like an umbrella, so that when they looked up, they saw an accumulation of snow caught in the green needles of the branches overhead. The ceiling of snow crystals and tree trunks trapped the heat, and although it was still chilly, the air was noticeably warmer inside the forest than outside it. "Wisps of pale green moss," wrote Kermit, "drooped from branch and twig. The rotting trunks of fallen trees lay on either side. There was an atmosphere of primeval melancholy and loneliness."

The forest the explorers were hiking through was thousands of years old. From the dead logs that Kermit admired grew young trees, each one an exact clone of its fallen mother. The spruce forest had remained genetically identical for its lifetime, an insulated community that persisted unchanged across the generations. Because the trees are identical clones, some scientists consider the spruce tree the oldest living organism on earth.

It was February, but Kermit could see an edge of bright green growing from the trees' dark-green needles. The new growth meant that it was spring and the travelers had been away from home for two months. Occasionally Kermit noticed a guide picking at the bright new needles, their texture soft and downy compared to the stiff structure of the fully grown vegetation, then popping the vivid green into their mouth and chewing. The young needles were packed with vitamin C, containing far more of the nutrient than a lemon or orange.

The next day the group continued their downward march, dropping some five thousand feet of elevation in two hours. Their knees ached from the slope of the precipitous, rocky trail, but they finally reached the valley below, where the Yalong River rushed swiftly beneath them. Two ferryboats sat in the water, each one made of two hollowed-out logs that had been fastened together.

Another group was waiting for the ferry, and Jack, Ted, and Kermit approached them to say hello. They were sunbathing by the water, and a few of the men had bravely decided to bathe in the icy, snowmelt waters. They were Tibetan nomads who made the high plateaus of the Himalayas their home, where the oxygen content in the air is less than half that at sea level and the average temperature is below freezing.

Jack struggled to translate, but with the help of the guides, the groups sat together, showing off their possessions with pride and warm interest. Ted displayed his rifle, which drew interest as he explained how to load the weapon and shoot. In return, the nomads passed a tsampa bowl to Ted. It was a small, round vessel, seven inches in diameter by six inches high, carved of wood with a lid that fit neatly on top. Lightweight and compact, it was a perfect companion for those who traveled frequently. Ted immediately wanted it. He was interested in everything and was always quick to purchase whatever items struck his fancy. He paid the equivalent of $1.20, then held the bowl happily in his hands.

The bowl was used for tsampa, which made up the bulk of the

Tibetan diet. It's a roasted and ground barley flour, usually eaten as a cereal with yak butter. Barley is one of the few crops that can survive the harsh conditions of the Tibetan plateau. Even though the Tibetans lived in different regions, spoke many dialects, and worshipped in diverse sects of Buddhism, tsampa connected them. The food staple would later be used as a rallying cry in the 1950s, when numerous articles in Tibetan newspapers called for the "tsampa eaters" to revolt against the Chinese. Tsampa became a unifying force and a central part of Tibetan identity.

The tsampa bowl was packed away and everyone prepared to cross the river. All the bags were piled in the middle while the people stood on the edges of the precarious craft. The boatmen used long staves of wood as their paddles. The mules couldn't ride on the delicate rafts and so had to be coaxed to swim across, a difficult endeavor that the guides struggled to achieve while Kermit, Ted, and Suydam decided to bathe in the river.

The icy water was not tempting—in fact, it took all their willpower just to dip in their toes, much less submerge themselves. Still, the glacier-fed

Crossing the river. Photograph by Suydam Cutting, 1929.

water had an undeniable appeal. It had been so long since their last bath that the grime and dirt felt like it was permanently crusted onto their skin. They stripped off their clothes but barely stayed in long enough to wet their hair before leaping out again.

On the other side, the explorers passed through a tiny village, just a collection of three houses, then began their upward climb once more. In the distance they could see where the trail was taking them, an agonizing route up and down the sides of a canyon that then rose steeply to a seventeen-thousand-foot mountain pass. "Looking forward from some pass such a maze of towering mountain ranges would fill the horizon," wrote Kermit, "that it would seem impossible that there should be a way to weave through them. The patient trail, however, wound along ridges, down into valleys, and around mountain shoulders, tracing a tortuous course."

End of the Caravan at 17,300 feet. Photograph by Suydam Cutting, 1929.

The next day they wandered into another village, where a temple built from gray stone stood out as the largest building. When they went inside, they noticed the floor was covered in small urns. Unbeknownst to the travelers, they had walked into a columbarium, a structure where funeral urns are held. They hadn't seen one before, and Ted was keen to explore. They found an interior flight of stairs that led into the main part of the temple, where a small group of people were gathered by a statue of Buddha near the altar. Ted, as usual, struggled with the Tibetan form of greeting: "It is most disconcerting to see a man smiling in the friendliest fashion and at the same time sticking out a very large tongue."

While Ted wasn't impressed with the looks of the temple, calling it "dirty and shabby," a shelf of holy books caught his eye. He signaled to Kermit and the two walked over to the volumes, "handsomely bound in stained wood and beautifully written in silver and gold wash. Everything was so ill-kept that Kermit and I thought we might be able to buy a couple of these volumes." They offered the equivalent of fifty dollars in local currency, closed the deal, and went happily on their way.

Jack told the Roosevelt brothers that there were rumors of large game in the countryside, so they decided to camp overnight outside the village and hunt the next day. In the morning, as they were setting out to search for any large mammals in the area, a ruckus broke out in the village. Everyone was in an uproar over the holy texts that the Roosevelts had bought the day before. Jack explained the situation: "The man you bought them from does not own them—they belong to the whole village." The townspeople had met the night before and told the would-be bookseller to either return the money and bring them back, or "leave the village forever."

Ted, not one to give up, tried to buy the books from the village instead. He was so used to money getting him what he wanted that he couldn't understand why the village refused to sell him the dusty old tomes. One of the guides explained that they contained the Tengyur, the sacred texts with words attributed to Buddha himself, but Ted still couldn't grasp the situation. "We tried our best to buy them," Ted explained, "without success. Though they were probably rarely opened and certainly not reverenced, they were regarded as a fetish. Of course, under the circumstances there was nothing to do but give them back, which we did with much regret."

The books that Ted dubbed a "fetish" were *terma*, or treasure texts, teachings that had once been hidden, waiting for the proper time to enlighten their followers. They were considered a continual source of inspiration, physical objects that formed a direct link with the mind. One of them was a guidebook for the afterlife, explaining how to liberate the soul following death; the Bardo Thodol Chenmo had recently been translated into English in 1927 under the title *The Tibetan Book of the Dead*.

"Are you oblivious to the sufferings of birth, old age, sickness, and death?" it asks. "There is no guarantee that you will survive, even past this very day!" The book teaches that life is cyclical, and death is not merely an ending but an essential part of how we live and the choices we make. "Life is continually arising, dwelling, ceasing, and arising," writes the Buddhist nun Pema Chödrön, "It's a cycle that goes on every day and continues to go on forever."

The Bardo Thodol Chenmo has gained popularity among cultures across the globe because it answers questions that directly address our fears of mortality. It tells us that our actions have consequences we cannot yet understand, and that death is not final, merely an opportunity to grow and become something deeper than we are. The book, meant to be read aloud, reveals that human lives have intrinsic meaning, beyond basic survival.

Ted was living his life with little regard for the consequences and with negligible thought as to where each action was leading. Yet every step on the trail was taking them deeper to an outcome that neither Ted nor his brother could predict, one in which death would mark the beginning of a greater transformation. Where Ted saw only dusty pages, there were messages that couldn't be purchased with money.

With the texts returned to the village, the brothers hiked to a mountain meadow and then scanned the edges of the trees in search of animals to collect for the museum. It was late in the morning, a poor time for such efforts, but they had to try.

Half of hunting is patience. It requires a person to be still and quiet for hours. Ted pulled out an old copy of *Great Expectations* while Kermit began reading *Personal narrative of a pilgrimage to el Medinah and Meccah* by Sir Richard F. Burton. The hours slowly passed. Then a sound came from the hillside. The Roosevelt brothers scooped up their rifles and set off quickly, trying to remain quiet but getting excited. A few villagers had joined them to help, since they knew this countryside best. They pointed to the opposite side of a ravine. There, in the bushes, a large animal emerged. It was a sambar, a species of deer designated *Rusa unicolor*. This one was massive, at least five hundred pounds, with a shaggy brown coat.

Ted yanked his rifle from its sling, steadied the gun on his shoulder, and took a shot. He knew the Field Museum would be eager for a sambar, and the expedition had garnered so few large mammals that he was desperate to obtain this one. A puff of dust could be seen on the hillside, and Ted knew he had shot too high. He aimed lower and fired again. This time he hit the animal. It fell for a moment but then, survival instinct kicking in, immediately popped up and began to trot away. Ted shot again, three times, and then ran to get closer. He finally finished the animal and saw, for the first time, that it was a doe.

Ted and Kermit measured the animal and then skinned the deer carefully so that the specimen would be useful for the museum. Each cut was made delicately so that the hide would be left as intact as possible for the scientists to examine back in the States and the taxidermists to stuff for the exhibit. Every bit of the animal would be used. The contents of the body cavity were removed entirely, and the meat butchered. The deer's blood was given to the villagers' dogs, and even the tear ducts were collected for traditional medicine.

In this way, the animal would live on after death. Its body would be used for comparative morphology, revealing, for the first time, the importance of the Asian continent in the evolution of giant deer. Its afterlife would stretch for a century, sparking wonder in generations of people who viewed the exhibit in Chicago. "They look like a cross of moose and capybara," a young girl would say ninety years later, staring at the same deer standing in the Field Museum that Ted had shot on a remote mountainside in the Himalayas.

Back at camp, the deer meat was cooked and buttered tea prepared. The Roosevelt brothers took a quick look at their maps before setting them aside. "We had secured the best maps obtainable," explained Ted, "but we found none to be of much use." They were hugging the northern edge of Yunnan Province, which Jack translated as South of the Clouds for the explorers. It truly felt like they were ensconced in fluffy, white puffs of cloud.

The mist obscured a large mountain that loomed in front of them. They had been watching the peak grow closer and closer as they hiked, aghast at its size. It was called Minya Konka, and its soaring white peak and the deep gorges surrounding it made the mountain look impossibly tall. Its remote location meant that no mountain climber had yet touched its pristine slopes to confirm its height. "There are those who claim that it rises thirty thousand feet and is the highest in the world," wrote Ted. Even this mountain, so majestic before them, had been left off their maps.

Ted and Kermit were becoming impatient. It felt as if they were hunting a beast in the clouds. The Forge of Arrows held all their hopes for finding a panda. It was there, in the bamboo forests near the village, where Ted and Kermit believed the purpose of their expedition lay, and where this trail would inevitably lead, if only they traverse it. "Let's lengthen our marches," Ted instructed, and although everyone was tired from the long days on the trail, they complied.

They hiked at night, the guides holding burning torches to light their way. "Their sputtering light gave the line a picturesque effect as it wound up and down the rocky path," as Suydam described it. The torches could only do so much. Ted had hoped the moonlight would guide them, but the sky was obscured by clouds, and it was impossible not to keep tripping over rocks on the trail. "The groping was abominable," wrote Ted. "In daylight and in fair weather it would not have been easy to pick one's way among the jagged boulders forming the trail, but in the dark with snow and ice coating the rocks it seemed impossible for the mules to get through whole in limb."

Hiking through snowy mountain trails. Photograph by Herbert Stevens, 1929.

The snow began picking up. It had been falling all afternoon but now it began coming down in thick, wet clumps that instantly soaked their hair and clothes. The torches were extinguished, and in the darkness, they kept stubbing their toes and twisting their ankles. The wind blasted between the mountain peaks, swirling the snow around them so that it felt like they were caught in a snow globe, barely able to see even a few feet ahead of them. The dark night was becoming dangerous indeed, and the guides called out for them to stop. They had to find a place to camp, but the question was where? In the blizzard-like conditions, there was no telling what the terrain around them looked like or where it was safe to pitch a tent.

Finally they bivouacked for the night on the side of the mountain, not bothering to pitch their tents, but simply crawling into their bedrolls, for those who had them, under a sheet of canvas. It was cold misery, with no fire to warm them and no relief to be found in their blankets, but at least they could rest. Yet as the hours passed, their old enemy, altitude sickness, caught up with them. This time, it wasn't just Ted, Kermit, Suydam, and Jack who were afflicted; many of the guides were also miserable. The high mountain pass, combined with their heavy exertions, had done them all in. They lay exposed to the cold and snow, drifting out of consciousness, only aroused by their bodies desperately gasping for air.

It's common for those new to altitude to sleep poorly at night. Sleep tends to be fragmented, lingering in the lightest sleep stages, a direct result of the lower partial pressure of oxygen in the air, which falls as barometric pressure lowers. Lying down makes things even worse. When upright, the diaphragm is held in its optimal position by gravity, maximizing the volume of each breath. But when lying down, the chest is slightly compressed. If you lie on your side, the heart will push against the lungs, compressing your airways and the lung parenchyma, clusters of tiny air sacs that allow for gaseous exchange. It may be just a small amount of compression, but even a slight reduction in the lungs' oxygen concentration can make a difference when you're already hovering on the edge.

Ted and Kermit vomited in the snow, too weak to wipe the sick from their mouths before lying back in their bedrolls. The brothers had never known pain like this before and their lungs burned intensely, as if a thousand needle pricks were stinging their ribs. Each inhale brought a fresh wave of pain, and Kermit arched his back in misery. The brothers grasped their heads in confusion, pulling their hair, as hallucinations overtook their brains. They didn't know where they were, what they were doing, or even their own names. Instead, in the dark of a starless sky, as they lay with tears streaming down their faces, the only answer to the horrific pain was death. It was their lowest moment, and what Ted would describe as the "worst night" of their lives. All they wanted was for it to cease, and in their altitude-sickness-addled brains, they begged for their breathing to stop.

Every minute that passed in the clouds brought them closer to their wish.

CHAPTER 7

FORGE OF ARROWS

The explorers woke to the gleam of a shining frozen waterfall. It curved over their heads dramatically, like a sculpture, the ice defying gravity as it pulled away from the rocky cliff in long, needle-shaped crystals. It seemed impossible they had slept under the frozen water all night oblivious to its beauty. The men and women arose from their cold beds and looked out over the land. The morning was clear, the earth tucked in under a heavy blanket of snow. The world looked new and fresh, untouched by humans.

It was hard to believe that they were alive. Their nighttime march had left them weak and pitiable, a group of men and women hobbling around camp so pathetically that no one coming across their party would take them for explorers. They looked like they were bound for the hospital. Or that they should be.

The first rays of morning sun set the waterfall ablaze in sparkling light. Kermit gazed at the ice crystals. He and his brother had died the night before, although, as *The Tibetan Book of the Dead* teaches, it was not an ending. Rather, the experience of extreme suffering had changed them in a real and perceptible way, letting them leave behind past failures and

Camping above the tree line. Photograph by Suydam Cutting, 1929.

insecurities and enter the next stage of their journey with a newfound humility. They awoke in the Himalayas as fragile as newborns, crawling out of their bedrolls and blinking in the brilliance. "We cling to such extreme moments," Peter Matthiessen writes, "in which we seem to die, yet are reborn."

There was reason for the explorers to find optimism in the dawn. As awful as the night march had been, it had brought them tantalizingly close to their destination. They were now a day's hike from Tatsienlu, or the Forge of Arrows, the likely home of the panda. They could see the town beneath them, "tucked deep down in the valley at the junction of two rivers." There was a Christian mission in the town, run by a Scottish couple, the Cunninghams, and it was at their house that the Roosevelt brothers had been invited to stay.

The Cunninghams were known in the area for their generosity. They gave medicine to the sick, ostensibly for free, although the real price was the gospel that was recited at the patient's bedside. Robert Cunningham had studied at the Manchester Royal Eye Hospital, taking a few classes in various eye diseases and their treatment. He called himself "Doctor" although he had not graduated from medical school; still, he had learned enough to be of some use to his patients. The missionary studied languages as well, although it took years before he was fluent. Until that time, a lama accompanied him on his visits around town, patiently repeating Cunningham's words, even his Christian prayers, though they were of a different religion than his own.

Jack, better than any of the other explorers, could appreciate the difficulty with languages in the region. A speaker of perfect Mandarin, Cantonese, Japanese, and Korean, as well as several other Asian dialects and some French and German, Jack could expertly mimic the accents of the people they met but found the skill to be of little use here. Although traveling within the borders of China, they were moving through a region so remote that not a single inhabitant spoke Mandarin or Cantonese. Such a diversity of ethnicities and languages meant that Jack was frequently lost and frustrated by his inability to translate for his team. He was left in an uncomfortable limbo, an interpreter who could not interpret, and his identity became submerged under the complexity of language and race.

Jack's descriptions of the people and places they encountered far surpassed those of his companions in detail and perspective. "These Tibetans are men of splendid physique and great strength, and are frequently more than six feet in height," wrote Jack. "Some are really handsome in a full-blooded masculine way. They wear fur caps and long loose coats like Russian blouses thrown carelessly off one shoulder and tied around the waist, blue or red trousers, and high boots of felt or skin reaching almost to the knees. A long sword, its hilt inlaid with bright bits of stones, is half concealed beneath their coats."

At last the explorers entered the Forge of Arrows, where they stood speechless for a moment amid the splendor. The town was perched on the side of a mountain, with an opulent palace occupied by the king and royal family, and surrounded by dozens of royal temples. It was clear that this was a seat of wealth and power unlike any other mountain town they had passed through. The streets were even paved with shining marble.

"The city, 5400 feet high, a thriving center for Chinese-Tibetan trade, has one of the most spectacular settings in the world," noted Suydam. "It has no tourists, however, for the approach is difficult from all sides, and only motives of profit would induce Chinese and Tibetans to come here. The first white man to describe the place arrived in 1868, and he has not had many successors.... All in all, the land is an enigma, and whatever the fate of the Chinese republic in the East, it will doubtless be centuries before the country can be subdued."

It had been weeks since the explorers took a proper bath, and they were excited that the Cunninghams possessed a luxurious wooden bathtub fitted with a stove underneath, so that the water became progressively warmer. The tub soon filled with black water as, one after another, they scrubbed their bodies and hair. They used bars of rice-water soap, and Kermit was so taken with the handmade product that he stuffed as many as he could fit in his sack. His skin felt new and clean, after layers of mountain dirt had been washed away, and he was certain the soap deserved the credit. The men didn't bother to shave; such grooming seemed pointless when it would all grow back soon enough—and they knew from experience that beards came in handy on cold, snowy mountainsides.

In town, Ted lingered over old bronzes and bits of ivory that were being sold out of stalls on the bridges, and soon his penchant for shopping was known far and wide. A steady stream of merchants began showing up at the Cunninghams' door, eager to show their wares to Ted. Many of them were lamas, and from the folds of their robes they showed off bells, drums, gongs, and horns of all shapes and sizes. Ted loved every minute of it. He never knew what treasures were about to show up at the doorstep and could happily spend an hour bargaining over an old bowl. A few items gave him pause, such as "bowls made from human skulls and the trumpets from thigh bones," and these he did not buy. Many other items, however, were thrown into the caravan's growing collection, much to the displeasure of the ten mules that had to carry it all over steep mountain passes.

The presence of so many salespeople, particularly the lamas, did not make Ted a pleasurable houseguest for the Cunninghams. Over dinner,

Robert Cunningham began to discuss the spiritual leaders, and the Roosevelt brothers were taken aback at his tone. Even though he could not survive in this remote area without them, Cunningham, like most Christian missionaries, did not respect the lamas, nor Tibetan Buddhism. They were "a low, demoralized, sensual, avaricious class," wrote one missionary in the region, "whose only care is to think out ways and means to get the possessions of the laity turned into the monasteries for their own use." The missionaries preferred the Chinese government which, because it was secular, presented no overt threat to Christianity.

One of the stories Cunningham told the Roosevelt brothers was that of Pere Davenat, a French missionary stationed at a post a week's travel from the Forge of Arrows. During the 1911 revolution and the collapse of the Qing dynasty, the lamas in this border region had risen up against the Chinese, declaring independence. The missionaries, due to their allegiance to the Chinese, were targeted and Davenat was taken, stripped of his clothing, and tied to a post in front of a temple. "Each lama," related Robert Cunningham, "on entering or leaving plucked a hair from his long beard." It took eleven days before the priest was rescued. "He never recovered," explained Cunningham.

It was difficult for the Roosevelts to reconcile the missionaries' experience with the lamas with their own. They were all foreigners here and dependent on the Tibetans. For Ted and Kermit, who had so recently been saved by the lamas during their trek in the Himalayas, and would certainly have perished without their help, it was impossible to join in the criticism. So it was a relief when Jack changed the subject.

"What do you think of the New China movements?" he asked Mr. Cunningham. It was a bold question. The New Culture movements in China were being led by young people who were protesting imperialism in Beijing and advocating for free elections, feminism, a greater role for science, and the modernization of journalism in China. Mr. Cunningham surprised Jack by agreeing with the progressive causes, and the two talked for some time. He was "sympathetic," as Jack later noted, "and understands the needs of the Chinese youth." It was a surprising turn of the conversation, and Ted and Kermit were happy to leave the divisive subject of lamas behind.

While the group chatted away, Suydam seemed interested only in the food. There were yak meat, vegetables, and rice, and he ate up happily, savoring the taste of food that hadn't been cooked over a campfire. He always loved eating, but the lack of food while traveling made the local meals seem even more delicious and noteworthy. While Ted and Kermit spent their time shopping for trinkets, Suydam quietly rated the food at each missionary's house in town. The French, he discovered, had a "fine, full-bodied" vintage of red wine that made the house his favorite, while the Americans were by far the worst. Neither alcohol, meat, nor tobacco was permitted in their home, so the visitors were served a "none too tasty stew... and regretted that this particular kind of piety excluded the good things of the world."

With their creature comforts attended to, it was now time to turn to the panda. The Roosevelts hired two local hunters to accompany them and, together with Suydam and Jack, set off early one morning for the mountains. They had their guns and supplies with them, but without their mules and the rest of the caravan, the travel felt light and easy. They raced up the paths like children, their feet beating the dirt through the North Lhasa gate. Ted turned around to admire it. It was built of two tall columns, and carved into the stone was a beautiful goddess riding on the back of a bumblebee. She was Chammo Lam Lha, and one of the hunters explained to Jack that she granted a safe journey to travelers. Around her lay flowers and other offerings. The hunters stopped before her and offered a prayer, as they always did before starting out. Then they continued up the road.

It was cold at these high altitudes, and a light snow began falling as they reached a small village some three hours north. The trail had narrowed and was now following a river. Several hot springs bubbled up along its edge and the smell of sulfur overwhelmed Kermit's nose as they passed. The pools were crowded, the first ones with men and others with women, all enjoying the warm, effervescent water.

Averse to questioning the naked people, the Roosevelts passed on by, but immediately greeted everyone else they came across. They constantly stopped and chatted with men and women in the villages and on the trail, showing their colored plates of the panda obtained from Paris. Every look was quizzical. No one had ever seen an animal like that before. It was discouraging, but there was nothing they could do but keep hiking onward. Nearly everyone they encountered, though they knew nothing of the panda, warned the Roosevelts of the deep snows that lay ahead in the valley, which would make the far reaches of the hunting grounds inaccessible. Not that it seemed to matter; there wasn't a panda living in the area in any case.

"This definitely discredited our original information," wrote Ted, and the group stopped that evening in a small village to discuss what to do. "We arranged to split," explained Ted. They flipped a coin and Suydam won. He and Jack would continue downstream while Ted and Kermit, who could never bear to separate, would go in the other direction. They set off, but the mood was no longer that of light, carefree children. Now they had work to do, and it seemed that everything they thought they knew about the panda's habitat was wrong.

Fog settled in like a blanket across the landscape, and navigation became challenging. Every direction looked like the same amorphous gray blob. The Roosevelt brothers stopped each person they came across, but no one had information on any mammal nearby, and the panda, even with its distinctive coat, had never been glimpsed by a single soul in the region. "We've clearly been led on a wild-goose chase," wrote Kermit unhappily.

Days passed and the snow began falling heavily. "It's not hunting weather," the local guides explained, but Ted and Kermit would not listen. They insisted on continuing their march, across the river and into the deep snowdrifts that lined the mountain valley. The new snow gave the forest a pristine beauty, making the green of the trees look richer. It was a "fairyland," said Kermit, although it was hard to enjoy the majesty of the area. Hiking in the deep snows was every bit as difficult as the hunters had warned, and it took four hours to get to the top of a small, nearby peak.

Meanwhile, not far away, Suydam had found something exciting: fresh bear tracks in the snow. His pulse raced. This was exactly what they had been waiting for. He followed the tracks cautiously, and then, at one spot where they were particularly distinct, stopped to examine them minutely. There were five toes on both the front and hind feet. Long claw marks stretched out from the front feet, close to the pads. The toes were arched distinctly and spread out in the snow with each step the bear took. Suydam got out his calipers and measured, his brain racing with excitement.

Could this be it? he wondered. *Have I found the panda?*

At the Field Museum, scientists had taught Suydam that bears were best identified not by sight but by their tracks. In the woods, your eyes can fool you. You will see a flash of fur, and adrenaline will flood your body, clouding your sight. Tracks, on the other hand, are steady and dependable—they exist, or they don't. Suydam wished he had his camera to document the tracks in the snow, but he had judged it too bulky to bring on the hunt. Instead, he sketched the paw prints in his notebook, then quietly followed them where they led into the forest.

As he traced the edges of the paw prints, he was certain he knew what species the bear was. While no one in Chicago, or most of the world for that matter, knew what a panda paw print might look like, the scientists had taught Suydam how to distinguish known bear tracks. What he was looking at now was a subspecies of *Ursus arctos,* the brown bear. The shape of the paw print, its measurements, and the way the claws stretched from the toe pads were all indicative of the grizzly bear subspecies. It might not be a panda, he reasoned, but finding any bear was cause for excitement, especially because it wasn't supposed to exist here at all. No grizzly bear had ever been described in China.

It wasn't outlandish to think that such an animal might prowl these mountains, as other species of brown bears had been identified elsewhere in Asia. Notorious among these was the Sankebetsu brown bear incident, which transpired in December 1915 in a new settlement on the island of Hokkaidō, Japan. The "worst bear attack in Japanese history" began simply enough, when a large brown bear sneaked into a family's house and raided the food stocks one night, escaping with a full stomach of corn. It was a subspecies of the brown bear called *Ursus arctos lasiotus*, whose size was massive, surpassed only by the Kodiak bear, *Ursus arctos middendorffi*, in Alaska, one of the largest bears in the world.

When the bear reappeared at the family's home some days later, a group of men shot at it, drawing blood but not killing the animal. It vanished into the forest, eluding hunters' attempts to track it. Then on December 9, the bear showed up at another family's home in the village, where it killed a baby and then mauled a woman, brutally killing her and dragging her body into the forest. The horror was not finished. The next night, the bear entered another home, where it killed a family of six, four children and their mother and father.

By December 13, the bear had caused so much physical destruction to the homes of the village that the land was declared "unhabitable for winter." A team of expert marksmen was dispatched from Sapporo, but they fired only at shadows, unable to track down the real thing. Finally a group of local hunters set out and succeeded in killing the bear on December 14, 1915. When they measured the animal, it weighed 750 pounds and stood 8.9 feet tall. All in all, nine people had been killed in one of the worst bear attacks in history, turning Sankebetsu into a ghost town.

It was a tale that Suydam knew well, legendary not only among the Japanese but repeated by hunters and explorers across the globe as evidence that some animals were simply too dangerous to live, with a propensity for human flesh. Although brown bears had never been described in China previously, it was not outlandish to think a similar species might be found throughout Asia.

The brown bear is designed for brutality. Its jaws can exert an astonishing twelve hundred psi, or pounds per square inch, enough force to crack a whale skull or split a bowling ball with a single snap. A solitary swipe of its massive claws, which can measure more than six inches, is enough to kill a person. Yet despite its ideal design for killing prey, the typical brown bear's diet is more than 90 percent vegetarian. The bear spends its summers picking berries and foraging before bedding down for a long winter's hibernation.

For all their peaceful existence, it was tales of aggressiveness and carnage that best defined bear stories among humans. While brown bears are more massive, black bears are considered just as dangerous. The mammal is a scavenger, easily spooked by sudden encounters, and known to kill humans that come across its path. Yet thanks to Ted and Kermit's famous father, black bears were also seen as adorable.

In 1902, President Roosevelt was hunting black bear, *Ursus americanus,* in the Mississippi Delta. His guide was Holt Collier, a famous African American hunter and former slave who reputedly had killed more than three thousand bears. While other members of the hunting party were encountering animals, the president was having little luck. Collier, wanting to help, took on the dangerous task of cornering a 235-pound black bear, lassoing a rope around the animal's neck, and tying it to a tree. When President Roosevelt was called over to shoot the animal, he flatly refused, calling it "unsportsmanlike."

The press immediately seized on the story, and the president's actions were soon national news, reported in papers across the country and satirized in political cartoons. In Brooklyn, candy store owners Rose and Morris Michtom were inspired by the tale and crafted a toy bear they named Theodore Roosevelt. They then wrote to the president, asking if they could sell the stuffed animals under the name Teddy's bears. The toy was an instant hit and by 1908 had become so popular that a minister in Michigan warned that "replacing dolls with toy bears would destroy the maternal instinct in little girls."

Whether stuffed with fluff or loose in the wild, neither black or brown bears can compare to the polar bear, *Ursus maritimus*, the largest carnivore on land, which can weigh more than 2,000 pounds and stand ten feet tall. The animal is by far the most lethal of the three, with a powerful bite and a paw swipe force of 1,800 pounds. An old adage goes, "If it's black, fight back; if it's brown, lie down; if it's white, say good night."

The dazzling coat of the polar bear was highly desired by European royalty in the thirteenth century, although its hair is not truly white. The fur is colorless, as clear as a water glass, with each follicle comprising a hollow core and covered in tiny bumps that act to scatter the light, projecting a white, luminescent glow. Just as the color of brown and black bears camouflages them in their environment, polar bears appear white to blend into the ice. Yet underneath their transparent fur is dark skin. The color is essential to life in the Arctic Circle, allowing them to soak up heat from the sun, while also offering protection from the harsh UV radiation caused by light bouncing off the reflective surface of the ice.

With this accumulated knowledge of bear species across the globe, scientists had hypothesized the behavior and habitat of the panda. What sort of strange environment did the animal occupy if its coat was truly black and white? And how would the panda's aggressiveness compare to its bear relatives? Whether it took more closely after the polar or the black bear, researchers expected the animal to be extraordinarily fierce, with a combination of bite force and paw swipe never measured on earth, and likely one of the most aggressive animals in the world.

While hunting animals known to kill humans can be frightening, for big game hunters it was part of the allure. The panda offered the ultimate challenge: a large mammal so elusive that no one had documented one, as deadly as a polar bear and as threatening as a black bear, occupying a habitat that no one could predict.

All day Suydam tracked the animal, long after the paw prints petered out in the snow, but there was nothing, not even a glimpse of the bear. He hiked until it was too dark to see, and even then he was loath to give up. Still, the tracks had given him hope that a bear might truly be found in the area.

Jack was hunting alongside Suydam, but he was focused on smaller mammals. While Suydam scoured the mountains unsuccessfully, Jack had collected an unusual squirrel species and several birds, all of which he "neatly prepared" for the museum. His talents were rapidly increasing. He could recognize different species, identify birdsong in the forest, track animals patiently, and hunt for them with a skill that far exceeded his experience. He was an easy companion too, funny and bright, and always eager to take on whatever work needed to be done. With Herbert many miles behind them, Jack was filling his boots, seeking the smaller animals and plants.

Meanwhile, Ted had finally found something. A group of bharal or blue sheep, *Pseudois nayaur,* were gathered in the distance. Their shaggy coats were gray in color with a slight blue sheen that gleamed when the sun hit it. The animal was native to the Himalayas and Ted knew it would be wanted by the museum. He followed the pack cautiously, trying to get within range of his rifle. There was a strange beauty to the animals and the way they ran across the mountain meadows on their light, sure hooves. Ted's footsteps seemed heavy by comparison.

With stalking skills honed from hunting with his father, Ted finally came into range. When he was about 150 feet off, he raised his rifle against his shoulder and fired. Two "fine rams" fell in the meadow. The rest of the animals scattered at the shocking, loud noise. Ted wasted no time in running to the animals and readying the specimens.

It was the first day of March, and Suydam, Ted, Kermit, and Jack regrouped to decide what to do next. Ted had bagged two blue sheep, Jack a collection of interesting birds and a squirrel, while Suydam and Kermit were empty-handed. The Roosevelt brothers were quick to praise Jack, and as he presented his specimens, he felt a burst of pride.

Jack's bright future as a scientist contrasted to their unfortunate present. As they sat around a fire, Ted and Kermit contemplated the dismal outlook. Every intelligence had led them here, and it had all been wrong. There was no telling where the panda skin the missionary had purchased for the American Museum of Natural History ten years earlier had originated. It could be from a forest ten miles away or a hundred. They were no closer to learning about the natural biology of the panda, and even more disheartening, they still hadn't met anyone who'd ever seen one.

For Suydam this was just another expedition. He would certainly undertake more. The man was useful and likely to be brought on other explorations. But for the Roosevelts, there was no telling. If they came back without the big prize they had set their sights on, the panda, it was likely that the Field Museum would not fund them again. Even this, their father's last legacy to them, felt like an ambition they could not realize.

They turned again to the empty spots in their maps. There was a mark on the map that they didn't want to consider, and which had never been included in their original plans, but now they wondered if they had a choice. Months earlier, back in New York, Roy Chapman Andrews had warned them not to enter this region, an isolated jungle that lay hundreds of miles east. "They're bloodthirsty savages," Chapman had cautioned of the people who lived there, although he had not entered the forest himself and knew almost nothing about the area's inhabitants. Chapman had even warned Eleanor, Ted's wife, telling her, "I only hope Ted and Kermit won't go into Lolo country. If they do, they may never get out alive. Those Lolos shoot Chinese on sight, and probably other foreigners too. They're primitive and wild and set on self-determination. It might be a very bad business."

The trail to this region was narrow, with not enough space for mules, which meant that their large contingent of guides could not accompany them. Without their helpers on the trail, the possibility of getting lost or attacked loomed large. It was a desperate play, the kind of undertaking that only a madman would dare seek out, but Ted and Kermit saw in this blank space of the map a possibility, a small hope that they might find their elusive bear.

CHAPTER 8

COMPLETE HEAVEN

The first time that Ted and Kermit entered the Forge of Arrows they had soared through the ornate gates in triumph. Months of difficult trekking had led them to what was supposed to be the home of the panda and their hopes were high. This time, as they passed through the gates back into town, they felt defeated. The local hunters stopped to say a quick prayer to Chammo Lam Lha, but Ted and Kermit hiked by the statue with barely a glance. It had taken extraordinary effort to get to this remote part of the world. They had nearly died in the Himalayas, and it was all for nothing. Not a single person here had even heard of the panda.

Months of planning had been necessary to bring their expedition to the Forge of Arrows, but the brothers now had mere days to plan the next stage of their journey. Even a week's delay could mean the expedition was waylaid by monsoon season. "The rainy season starts in April," a young man had told them in December, "and if you run into it, it will be impossible to hunt and travel."

The Roosevelt brothers traced a line through the empty spots on the map and hoped they knew what they were getting into. They decided to head north, through Muping, where they hoped that dense bamboo forests tucked into steep, icy mountains might hide elusive creatures. In case the panda wasn't lurking there, they traced another line. This one was the route they didn't want to take. It led south to Yachow, known today as Ya'an, and then to a remote, mountainous province hundreds of miles east occupied by an ethnic minority known as the Yi people, and which the ethnic Chinese called "Lolo land." It was a part of the world to which Roy Chapman Andrews, although known as an intrepid explorer, would not travel because of its dangers, and that the Christian missionaries in Tatsienlu, although keen to take risks in their pursuit of spreading the gospel, would not enter. "The land of savages," one of the missionaries explained to the Roosevelt brothers. "Don't go there."

Putting the map away, they next selected a few favorite guides to accompany them, the ones they trusted most highly. The only thing left to do was to pack light, bringing just the necessities with them on the trail. Since they wouldn't have the mules, every item had to be chosen carefully, as the weight would slow them down. This meant that all their specimens, and Ted's many treasures and knickknacks, would have to travel the long way to Yachow, where Ted, Kermit, and Suydam could meet them. It also meant that they needed someone of the party to travel with the mules and remaining guides.

Ted broke the news to Jack. "We're going to meet you in Yachow," he explained. It would be only a month's separation, Ted stressed, but Jack knew how much could happen in thirty days. He thought of Herbert, still wandering somewhere behind them. They were supposed to meet the scientist here, but instead they were going on without him. Jack wished he was more useful, and that his language skills had made him indispensable to the Roosevelts, but he consoled himself that he was now in charge of the caravan. At least the brothers trusted him.

Herbert presented another problem. While the Roosevelt brothers had faith that Jack would meet them where and when they arranged, there was no telling exactly where Herbert was, or how far he might be behind them. With no direct means of communication, they had to simply leave word for Herbert about their plans and write him a note detailing a different location and time to rendezvous. Everything was in chaos, and they had no idea if he'd receive the new instructions.

Herbert was, in fact, months behind them. Left to his own wanderings, the scientist was weaving this way and that through the mountains, taking time to appreciate the diverse amount of plant life and birds around him. He was still outside of Muli, and the going was rough. With only one guide, it was easy to become lost. "I was in a quandary," he wrote, "to find at one point we were actually going South, when the trend of our direction should lie almost due North." Yet as varied as his trekking had become, he was still headed for the Forge of Arrows, where he expected to meet the Roosevelt brothers.

After much hurried effort, the Roosevelts and Suydam were able to arrange their belongings, set their course, and select the right guides. They set off early on the morning of March 6, after a quick good-bye to Jack. "The lazy indolent feeling of Spring was in the air," wrote Kermit, "and after the long weeks in the high altitudes of biting cold we all felt relaxed and uninclined to effort." It was hard to be in a hurry when you were still unsure of your destination.

The group sauntered along the trail, admiring the flowers and trees that had burst into bloom seemingly overnight. Orchards filled the meadows on this side of Tatsienlu and "the pear, peach, and apricot trees wore a delicate radiance of buds." The air felt "balmy," and the trail was filled with delicate flowers, none of which Ted and Kermit knew the names for. If Herbert had accompanied them, he would have insisted on lingering. The valley was a gift to botanists, filled with some of the most beautiful ornamental flowering plants in the world, many of them species entirely unknown to Europe or the Americas. Intrepid botanists had barely scraped the surface of unusual species in Western China and Tibet, where they found flowers that no Westerner had ever glimpsed before, including pure-white lilies, delicate wild roses, azaleas, rhododendrons, and the greatest diversity of orchid species on the globe. Flowers that Herbert could have spent hours examining, Ted and Kermit passed with only a glance.

COMPLETE HEAVEN

China is home to an astonishing 10 percent of the world's flowering plant species, more than three times as many as in the United States and second only to Brazil. The varied landscape that the explorers were hiking through was the secret behind China's biodiversity. The lowlands were warm and tropical, while the harsh mountains, rising high in the Himalayas, gave way to vast deserts and then to the Tibetan Plateau, the highest flat upland in the world. The plateau is sometimes referred to as a "third pole" because outside of the North and South Poles, it contains the largest reserve of fresh water.

The other secret to biodiversity in China is that the land was spared from the dramatic ecological turnover that occurred in North America and Europe during the last glacial period, ten thousand years ago. While ice sheets covered massive swaths of the Northern Hemisphere and smaller sections of the Southern Hemisphere, ancient plant lineages were able to persist in China, evolving into new species within an intensely diverse plant community.

The botanists who traveled to Western China entered a preserved forest, which some described as akin to a "museum" because of the extent of ancient and new species of flowering plants. Like a water cup filled to the brim, the flowers crowded the mountain meadows that led down from the Himalayas and the Tibetan Plateau. The Arnold Arboretum in Boston, a botanical garden run by Harvard University, was particularly intent on collecting samples from this region to fill its park. The university had funded several expeditions, including those by Joseph Rock, who wrote of "seeking strange flowers, in the far reaches of the world." The Beijing lilac, *Syringa pekinensis*, that Rock brought to Massachusetts as a seed still unfolds its creamy yellow blossoms in the outskirts of Boston today.

Rock had not traveled this far, so much of the plant life the Roosevelts passed had not yet been described by Westerners. Kermit took samples here and there, grabbing at any plant that struck him as interesting, but his descriptions were mere ramblings compared to Herbert's expert field notes. Still, they weren't certain if the scientist would ever get to these far-flung meadows, so the Roosevelt brothers did their best.

It wasn't always easy to collect plants. Almost immediately they noticed how crowded the trail was. Men and women hiked alongside them, carrying impossibly heavy loads. Tea was pouring into Tibet, where the drink was immensely popular, carried on the backs of tea porters who had been trained for the profession since childhood. As early as six or seven, children were taught to carry packs. Those packs became progressively heavier as the children grew older, until they were able to carry hundreds of pounds on their shoulders along steep mountain trails.

"They are of all ages and both sexes," wrote Kermit. "The day's march is short, rarely exceeding ten miles, but although the trail here on the main road was not too bad, after it turned off across the mountains it changed radically in character, and was difficult to negotiate without any burden.... The tea is packed in slender rolls of matting three and a half to four feet long and a load will measure as high as six feet, and all roadside accommodations are built with these measurements in mind."

Suydam was struck by the size of some of the loads the men and women carried. "It is nothing uncommon," he wrote, "to see a grown man carrying a 350-pound burden over a bad trail for eight miles a day." One load in particular was so massive that Suydam stopped to grab his camera and take a picture. The rectangular packs ran high over the man's head, and Suydam guessed the load to contain over 400 pounds of tea. He took the picture quickly, unsure if anyone would believe him without proof.

A large tea load carried on the trail. Photograph by Suydam Cutting, 1929.

As the explorers trudged along their trail, their own packs felt uncomfortably heavy without the benefit of the mules they were used to. They were carrying just a month's supplies, but every pound felt like ten as it bumped against their backs while the trail dropped six thousand feet in elevation. It was nothing compared to their guides, who were loaded down with weighty packs. They were now being paid by the pound, which Ted called a "most satisfactory method." It certainly was for him; his own pack was one of the lightest since he, through the Field Museum, was the one paying. The guides whose backpacks were the heaviest were paid the most, but among themselves, the men and women took turns to keep their salaries equitable.

The trail was filled with guides, tea porters, and even porters who toted men along the trail. Ted watched as one young Chinese official refused not only to carry anything himself, but even to walk on his own two feet. He hired a tea porter to carry him, piggyback-style, down the rough mountain path. The young man rested his head on the porter's hair, his arms dangling on either side as the porter beneath him walked. It made for a disturbing sight, but the Chinese official was undoubtedly lighter than many tea loads.

They had started the day with their feet in snow, and by the afternoon a warm breeze rustled their hair. That evening they stopped at a small inn, where

Chinese man being carried by a Nashi porter in a backpack-style chair, slightly different than being carried piggyback style. Photograph by Suydam Cutting, 1929.

Suydam had a chance to observe the tea porters up close. A wooden bench lined the hallway as one entered the inn, its height perfectly proportioned for a person unloading their tea for the day. A line of packs was already sitting in a straight row, left by their owners for the night.

Inside, the men and women sat in exhaustion. Suydam noticed that as strong as the tea porters were, their physical condition was concerning. Their leg muscles were massive, but otherwise their bodies were emaciated, as if they hadn't been getting enough calories to sustain their high level of activity, and so the muscle and fat were carved straight out of the torso and arms. They all had "deep circles under their eyes," but it was impossible to tell if this was a result of fatigue or opium, a drug that nearly every tea porter seemed to be addicted to.

The acrid smell of opium, as sharp as ammonia, filled the rooms of the inn, where the porters sat with pipes in their mouths. During the day, as they couldn't stop to smoke the drug, they swallowed pellets of it, the drug bitter but far more potent. Although opium was technically illegal, it was everywhere. In Buddhist temples, the black tar was smeared across altars as an offering, and Ted noticed that even government officials smoked the drug. "It's like prohibition in the United States," Jack wisely observed. Among the tea porters, the ramifications seemed to be the greatest. "What the physical drudgery begins," argued Suydam, "the opium finishes. Heart ailments, varicose veins, and undernourishment make an early end of them."

The next night, in a tiny village, Ted and Kermit spied an old takin skin, an unusual species of goat, hanging in a window. They went inside and, aided by one of their guides, who spoke English and could translate for them, learned that it had been shot by an old man who lived way out in the country but who happened to be in town. They found the old hunter and began questioning him eagerly. "Have you seen this?" they asked as they showed him the plate of the panda. The man looked at the picture of the black-and-white bear, and then nodded.

"I trapped one in a pitfall beyond Muping," the old man explained. "I still have the skin."

A tingle ran through Ted and Kermit's bodies. This was the first time in their travels that someone had claimed to have hunted a panda, and they began to question him, wanting to know everything about the animal's habitat, diet, and behavior. The old man could tell them little, for it had been many years ago and he hadn't seen one since, but he did offer his opinion that, because of the time of year he'd captured the animal, it was not likely to hibernate, a point of contention among biologists.

In 1908, a botanist named Ernest Wilson, working in China, spotted "large heaps of dung" from a bear that he believed came from a panda. He was collecting specimens in the Min River valley, and although he never saw the creature, he postulated about its behavior and diet from the scat he found. He guessed that the animal, like most bears, hibernated. It was a reasonable assumption. Without any other information about the animal, it seemed likely that the panda resembled other known bears in terms of behavior.

Ted and Kermit arranged for the old man to journey home, a day's march away, and meet them at their next destination, Tienchuan, translated for Ted and Kermit as Complete Heaven. They might never be able to find a panda, much less shoot one, but at least they could purchase a skin from a local hunter. The more they questioned the man, the further convinced they were that he was speaking the truth.

It took twelve days for the small party to reach Complete Heaven. They stopped to hunt only once, and, as usual, came up empty-handed. Immediately, the town surprised them. "It seemed incongruous to find such a large town having no connections with the outside world," wrote Ted, "either by land or water. Everything that was brought in or taken out traveled on man's back." He did not mention women, but they were everywhere, carrying massive loads of tea, and now even coal, which he had not seen before, from a local mine. Unlike the other towns they had traveled through, there were no mules or horses here. The village was completely isolated: no roads connected it to any other city. The only way to get in or out of Complete Heaven was by foot trail.

Complete Heaven comprised an intricate maze of streets, with beautiful carved arches spanning many of the main roads. There were several temples, marked with large stone columns for worshippers to enter their lush courtyards, filled with fruit trees and flowers. Dove trees stood tall on the corners of streets, the white flowers in bloom and looking like handkerchiefs stuck between the branches. Everything might have been lovely, a complete heaven indeed, but instead the opposite was true. It was as far from paradise as Ted and Kermit could imagine.

The brothers had seen illness before. It was common to see goiters, an enlargement of the thyroid caused by iodine deficiency, in the small towns of China. Here in Complete Heaven, however, disease was everywhere. Men and women hobbled along the streets, clearly in pain from an infection that Ted and Kermit could not guess, while others sat in silent misery along the street corners. There were many blind residents, more so than in other towns, and even goiter, which they had seen so often, was different here. It was a "particularly aggravated form," noted Kermit of the massive, uncomfortable growths that hung from the necks of men and women. As the group turned a corner they saw a child lying in the street, near a temple gate. Ted and Kermit approached, fear prickling their skin, but quickly saw that the boy was beyond hope. He lay there dead, with flies buzzing on his eyelids. It was terrifying.

When they'd initially picked Complete Heaven to pass through, the name had seemed a cheerful omen, but now they wanted nothing more than to leave. Even staying one night seemed dangerous. The next morning, they rose early. It was an almost desperate feeling as they packed everything up and prepared to flee. The guides shouldered their heavy packs without complaint, and they were off.

It was March 1929, marking four months since any of them had had news of their families. Months of isolation still stretched ahead of them, but at least they had escaped Complete Heaven unharmed. As they left, they realized they hadn't waited for the old man to come with the panda skin. They had promised him payment, and they really needed that skin, but it couldn't be helped. They had to get out of the town.

On the outskirts of town, Ted caught sight of the old man. The brothers eagerly made their way to him. He had the skin, and the guides quickly negotiated payment for it. The specimen had been dried and stretched and the Roosevelts began to examine it. They could hardly believe their eyes. The black-and-white color was extraordinary; there was no mistaking this skin for that of any other animal in the world. Still, the specimen was old, and Ted was worried the museum might not be able to mount it, especially as the skin had been ripped in a few places. Now that they knew that pandas could be found nearby, they were anxious to get going. It was an eight-day march to Muping, and as they hit the rocky trail, Complete Heaven no longer seemed like such a woeful town.

Meanwhile, Jack was experiencing the tense anxieties of leadership. For the first time, he was in charge of everything. The caravan, filled with various animal specimens, camping equipment, and numerous treasures purchased by the Roosevelts, was an enormous responsibility, and Jack spoke with the guides to make sure that their route was clear. He did not want to become a Herbert, left behind because he was too slow. He had the date written in his diary, and he was determined not to be late.

Jack was near the head of the caravan, walking with the guides, when the trail rounded a bend. Suddenly they were surrounded by twenty bandits, fanned out across the road, pointing their rifles at the explorers. Jack's heartbeat pumped irregularly as adrenaline flooded his system. He paused for only a moment and then, with his pulse racing, he did the unthinkable. He ignored them. The guides matched his bravado, and the group walked slowly by the cluster of bandits, saying nothing, merely keeping a slow and steady pace. Jack had a rifle slung over this shoulder, prepared for this very moment, but he did not use it. Instead, he kept his eyes ahead and did not engage with the robbers.

Jack knew he and his guides were outnumbered in terms of weaponry. Further, his command of the language was shaky, and, being young and relatively small in stature, his appearance was not imposing. More or less ignoring them, he figured, would at least surprise the robbers. They stood still, their guns pointed at Jack and his companions, but they did not try to stop them or even say a word. When the caravan had passed them and the robbers were out of rifle range, Jack trembled with relief.

He had pulled it off, and no one was more surprised than he that it had worked. The next encounter with thieves, however, would not be as easy.

The caravan was right on schedule, reaching the halfway point of their monthlong trek to Yachow, when disaster struck. Jack was out ahead with the guides again, his rifle securely by his side, when he sensed trouble ahead. He heard noises, maybe even shouting, he couldn't be sure. He halted the group, told everyone else to wait for him, then took off with two of the guides to see what lay ahead. They left the dusty trail and made their way through the dense, shrubby vegetation on the left side, careful to move as quietly and quickly as possible. Coming to a small rise, they crawled on their stomachs in the dirt, then peeked their heads down at the trail below. Perched on either side of the path were dozens of men, at least fifty, armed with rifles. A short way ahead of them, a woman was crying out for help. Her cries were obviously false. She was the bait, meant to lure caravans into the clutches of the bandits as they waited on either side of the bushes.

Silence was even more important now. Jack knew they could not be discovered, or they would be immediately overtaken and likely killed. They picked through the low-lying bushes like ballerinas toe-stepping across a stage, each movement deliberate. When Jack accidentally rustled some branches, he felt a shock of terror run through his body. He kept still, his own hand planted over his mouth to ensure he didn't make a sound. Besides a few melodramatic whimpers from the woman sitting on the trail, all was quiet from the other side of the hill. The robbers were striving to be as careful and silent as Jack and his group were.

With the utmost care, Jack and the guides made their way back to the caravan, where they quickly explained the situation and then turned around, heading back to the small town they had just stayed the night in. Jack knew there was no way they would be able to pass the robbers on the trail; they resembled a small army more than a gang of bandits. This time, walking slowly and ignoring the enemy would not help.

Back in the relative safety of the country inn, Jack considered what to do. Signs were posted around the town that warned of the penalty of robbery. *All proved bandits must have their heads cut off without trial,* they read. Jack's options, he was aware, were limited. It was possible he could negotiate with the bandits, perhaps pay them money for safe passage, but how could any meeting be safely arranged or any agreement be trusted? Jack asked the guides for their advice, and they were clear: he needed the army. These robbers, they explained, didn't care who they stole from or what they took. If the forces were that big, the only way to get through was to match them gun for gun. And the only people with that much firepower were the Chinese military.

While traveling along the border between Tibet and China, the travelers had already seen evidence of skirmishes between the two countries, but in this area, the conflict was stark. Here, warlords ruled the region, and they had no intention of giving up their power to the new Chinese government. China was caught between political futures. The 1911 revolution had ended a millennium of imperial rule, but the new Republic of China had yet to solidify power across the vast nation. In the chaos of change, local warlords, or *junfa*, seized large swaths of the countryside, shaping the land with their violence. Historian Andrew Nathan has described this period as "the darkest corner in twentieth-century Chinese history." Now Jack and his caravan were caught in the shadows.

For Jack, who still resented the damage China's nationalist army had inflicted on his family's fortunes, the choice was difficult but clear. He needed soldiers to clear the path. In the small town, still many miles and a fifteen-day march to Yachow, where he was supposed to meet the Roosevelts, he waited impatiently. The days ticked off, one by one, and Jack grew more agitated. He hoped the Roosevelt brothers would wait for him, but he knew from experience that they hated postponing their travel. To the local officials, Jack stressed the urgency of his safe passage repeatedly, hoping he made his pleas clear.

Jack had no set idea how many soldiers would be needed to clear the road, so he was amazed when fifty armed men showed up ready to fight. They had been drawn from posts all over the region. The fighting lasted three days, but when it was done, Jack could finally move on with his caravan. It was a strange, liberating feeling to have the army working for him for a change.

Meanwhile, the Roosevelt brothers and Suydam were just setting foot in a new town. "Muping," wrote Ted, "is one of those strange little semi-independent feudal kingdoms that fringe Tibet." As the expedition had skirted the border between Tibet and China, they had passed through mostly autonomous regions, ruled by royal Tibetan families. Here in Muping, however, the situation was altered. A year earlier, the last prince had died, leaving his wife and daughter. Without a son, there was no one to reign over the kingdom. A match was made with a prince from the Forge of Arrows, who came to Muping and married the widow. The outsider, however, seen as an easy target, had been killed by the Chinese, and a Chinese magistrate sent to rule the area. Fearing for their lives, the wife and young daughter lived in "poverty and seclusion" in the town.

When Ted and Kermit met with the Chinese magistrate, they were "astonished" at his lack of information. He did not know the name of the river that flowed nearby, nor where the trails led to out of town. He explained his ignorance by telling the brothers that he "had only been here one year," but Ted still shook his head. The man in charge of Muping apparently knew nothing about it. When the magistrate told him that pandas were nearby, although "shot only rarely," Ted and Kermit were unimpressed. This wasn't a person who could be relied on for accurate information, they surmised. Still, it was pandas they were here for, and they would do their best to find one.

After a bath, the Roosevelt brothers strolled the town. They gazed up at what had been the king's palace before the Chinese had seized the town, now partially collapsed, with two stone lions defending the ruins. Evidence of the Chinese overthrow was not confined to the palace. The largest temple in town was not Tibetan, as in so many places they had passed through, but Chinese Buddhist.

While there were similarities in the teachings of both religions, the absence of the lamas could be felt profoundly. Kermit had only to look at his coat, the patches of the holy yellow fabric still prominent, to remember how important the spiritual leaders had been to their journey thus far. The Chinese government had killed or evicted every lama in town and was determined to wrest power from the Tibetan leaders, whom they accused of "barbarism."

Buddhism had originally come to China from India during the Han dynasty, two thousand years ago. The Buddhist missionaries who traveled to China translated their texts and adapted their religion to fit in with Chinese culture and tradition. Their success fueled the hopes of Christian missionaries who arrived in the late 1800s wishing for the same success, but the reaction to the Westerners was not the same. The Christian missionaries had been allowed into China as a condition of the ending of the Opium Wars, and their presence was seen as an extension of Western imperialism. Still, their missions scattered across the remote Chinese countryside were convenient for the Roosevelt brothers, who stayed in the homes and enjoyed their bathtubs.

Suydam and the Roosevelt brothers were dining at the Chinese magistrate's house that evening. Before they left the missionary house, visitors were announced. Ted and Kermit came to the front room where they saw a middle-aged woman and her daughter, a girl of about eleven years old, waiting for them. The woman introduced herself. She was the widow of the Tibetan prince who had once ruled this region and who had been assassinated. She lived in hiding with her daughter. "Please help us," she begged through one of the Roosevelt's guides, acting as interpreter. She pressed a gift of dried meat into their hands, an item of great value to her, then tried to explain the situation. She believed the Chinese government was looking for them, and when they had hunted down the last remaining members of the royal family, she was certain they would kill her and her daughter. She was a grave threat to the Chinese, she explained, because whoever she married would be the true ruler of this region.

The Roosevelts had passed through several autonomous or semiautonomous kingdoms on their trek and it was impossible not to have sympathy for the poor woman. They were reminded of Muli and the guardian of the eastern border. In the House of the Prince they had been treated with tremendous kindness, and a friendship had formed. Yet what was the future of such kingdoms, clinging to the remote borders of the Chinese wilderness? Would the rowdy three-year-old in Muli one day share the same fate as this woman's husband?

The situation was impossible. "I am persuaded that their fear is groundless," wrote Ted, "but it was very real to them." Even if he believed them, there was nothing he could do. He had not the power or the experience in the region necessary to slow the onward march of the nationalist government across Western China. All he could do was talk to the magistrate, a man who was not inclined to listen to his opinions or those of any other outsider. After all, Ted was a Westerner who knew nothing about politics in the area. The Roosevelt brothers gave their visitors a silk scarf as a gift, and then said good-bye, as they now had to rush to the very house that the widow and her daughter most feared in the city. As the pair left, Ted heard them stop by the kitchen, where they asked for an empty tin can. He did not know what they needed it for, but pity filled his heart.

Nothing ages the body and the mind quite so fast as travel. Over the course of sixteen weeks, wrinkles around Ted's and Kermit's eyes had deepened, gray tinged the edges of their whiskers, and their opinions, although not as easily glimpsed, had been altered too. They were already different men than those who had first trod the trail four months ago, yet there were still many miles to go. Their ragged appearance was a frequent joke among the guides. "Are you sure you're a Christian?" one asked Ted jovially. With their wiry beards and patched clothing, the Roosevelt brothers looked nothing like the sons of an important man.

Arriving at the magistrate's house, Suydam, Ted, and Kermit were warmly welcomed and seated around a table where glasses of rice wine were poured. Introductions were made with the other guests, and the explorers learned they were joining a particularly fierce gathering of dinner guests. There was the head of a group of local hunters, the leader of the local Chinese militia, and a troop commander, in addition to the magistrate himself. Within these walls were all the men who had seized the kingdom, killed the ruler, and changed the fate of Muping.

The other guests were fast to make conversation, peppering Ted and Kermit with questions. "What does the United States think of Dr. Sun Yat-sen and the Chinese republic?" "Will the United States be able to stop another European war?" "Why are Japan and England unfriendly to China?" The questions were constant and the conversation lively as the various dishes were served to the Roosevelts.

The dinner seemed to be the "usual affair." There were vegetables and chopped meats on large platters. Suydam, Ted, and Kermit had been served all kinds of food on their journey and were not picky eaters. They knew that part of being a good guest was partaking of what was offered to you without complaint or a trace of disgust. But they were not prepared for the next course to be served. With ceremony, a large platter was put on the table. This was clearly a dish that the magistrate was proud to be able to offer to his guests. With a start, Kermit realized that monkey paws were sitting atop a bed of vegetables and broth. The humanlike fingers reached up from the table toward the sky. A murmur ran across the table; from the reaction of the other guests, this was clearly a rare treat and a meal to be savored. Ted and Kermit gave each other only a quick glance before turning to their plates. They held their chopsticks aloft for a moment, and then dug in.

CHAPTER 9

KINGDOM OF THE GOLDEN MONKEY

The old man was hunched over his work, a knife in his hand as he carved a ladle out of solid beechwood. Kermit and Ted approached him gently, not wanting to startle the old fellow. The Roosevelt brothers were a few days outside Muping and had been hiking through thick underbrush for an hour. They were surprised to see anyone out here, miles from any town, and stopped to say hello.

That morning Suydam had agreed to explore the hills to the south while Ted and Kermit hiked north up the valley where a thick bamboo jungle awaited them. They were escorted by a group of local hunters who knew the area intimately and were prepared to delve deep into the forest. Jack was still escorting the caravan east while Herbert trailed along behind them to the west, moving at his own leisurely pace. The four of them were stretching in every possible direction, desperate to take what they could from the vast Chinese wilderness. Nothing could match the panda, but they needed to bring home something notable.

As Ted and Kermit chatted with the old man, they asked if he'd seen any animals in these woods. The man nodded. "Just a few hours ago," he told them, "while wandering in the jungle I saw a troop of monkeys." Kermit peppered him with questions: "What did they look like? How many were there? Have you seen them before?" But the old man shook his head. They were unknown to him, but a strange yellow color that he would not soon forget. Kermit and Ted stared at each other. *Could it be?* they thought. *Were these the golden monkeys?*

Making more noise than was wise, they plunged into the thick undergrowth of the forest. They had heard of the golden monkey, but little was known about the elusive creature. Its description was mythical: a monkey with rich golden fur and a bright-blue face. How could such a distinctive animal be real?

The golden monkey had long been desired by Ted and Kermit. Before leaving New York, the Roosevelt brothers had stopped by the American Museum of Natural History to brush up on a few scientific techniques for their expedition. They had been chatting with journalists in the hallway when a scientist came up to them in a rush of excitement, proclaiming, "Colonel, how I envy you going where those beautiful monkeys are!"

One reporter, sensing a story, turned to the scientist and asked, "What kind of monkeys?"

"Why the golden snub-nosed monkeys, of course!" the man replied, and then rushed off, muttering to himself.

"Do they worship monkeys there?" the reporter asked, turning to Ted.

"Not there but here," Ted replied lightheartedly.

There was some truth to the jest. The golden monkeys were highly desired by museums across the world, as no institution held a complete group of them. They had first been described in 1897 by the French scientist Henri Milne-Edwards, who had received a specimen from a French missionary in China. Despite a few scattered specimens held in museums, naturalists wanted a family group of the monkeys as they believed that social dynamics made up a critical part of their evolution and wanted to compare the rare monkey to similar species.

The golden snub-nosed monkey, *Rhinopithecus roxellana,* endures the coldest temperatures of any nonhuman primate in the world. Their thick golden coats, ranging from a light yellow to a rust brown, help them withstand freezing temperatures on the Tibetan Plateau and a habitat that is covered in snow for four months a year. They sleep in large groups at night, high in the treetops, cuddled together for warmth. More than 95 percent of their time is spent in the canopies of trees, moving down from the safety of their branches only to forage or travel.

The kingdom of the golden monkey is ruled by a fellowship between animals and plants. The forest feeds the monkeys, but they in turn sustain it by spreading plants across the jungle. The golden monkeys' feces are full of seeds, each of which has a significantly greater chance of germination than if it hadn't been consumed at all. The monkeys play a major role in their ecosystem, regenerating the forest simply by their presence.

The Roosevelts moved through the bamboo jungle eagerly, moving up a steep slope. They could make out nothing in the distance, so dense was the tangle of vines and "blanket-like foliage" around them. Suddenly a yellow streak caught Ted's eye. It was moving fast through the branches to his right. He held up his shotgun but with no hope of hitting his target; he could see nothing through the trees. In desperation, he fired a shot and, much to his surprise, heard a loud thud. Ted rushed to the fallen animal, taken aback by its size. Its fur was long and golden and its belly an intense orange color. Sure enough, its face was a light sky blue. Even in death it was beautiful.

Ted heard another rustle and began shooting; then Kermit behind him started firing as well. Soon the jungle reverberated with the sound of gunshots. The noise was so loud that it was as if a war were being waged on nature itself, and perhaps it was. "It sounded like a miniature battle," Ted said, "as we fired at half-seen shapes flitting through the tree-tops." When it was all done, Ted and Kermit looked down proudly and "counted their trophies." They had killed an entire family of nine golden snub-nosed monkeys. Such a thing had never been done before. In their excitement they felt the rush of unveiling a rare species. They could not yet know that the monkeys they'd killed would, like so many other museum exhibits, become symbolic of the last of their kind.

Before the animals became trinkets in natural history museums, they were sources of wonder, especially for the Roosevelts. The expedition had meaning to the brothers that was difficult for them to explain to others. Money, fame, and success might make the world turn, but the Roosevelts also possessed a deep love for wild places. Hiking through the jungle was igniting their sense of curiosity. It reminded them of their youth and weeks spent with their father in the woods. The deaths of the golden monkeys jerked them back to the reality of their daily lives. It was a reminder that as beautiful as these animals were, they were worth their considerable weight in money and prestige.

Ted and Kermit were seeking out their guides for assistance in carrying the animals out of the jungle when they heard shooting below them. They had thought that they were alone hunting in the valley. Among their party, the Roosevelt brothers had been insistent that only they would be shooting game, and what's more, they wouldn't permit the use of hunting dogs. This had been met with confusion among the local hunters accompanying them, who had assumed that Ted and Kermit would rest comfortably at camp while they went out and did the work.

The lead hunter had argued with the Roosevelts, explaining that no important men ever shot their own animals, but Ted was adamant. The expedition was ostensibly for obtaining animal specimens for the Field Museum, but for the Roosevelt brothers, hunting represented more than merely the outcome. There was a glory in the hunt itself that was their father's legacy, and which they could not hand over to others, even those with more experience and thus a better chance of success than themselves.

While most scientific expeditions of the era were large, highly organized affairs, accompanied by fleets of motor vehicles and amply provided support staff, the Roosevelts traveled more like wanderers or hobbyists. The Roosevelt brothers possessed a humanity in their worldview that was starkly different. Their group was small, contained, but highly passionate.

When the conversation grew increasingly argumentative, Ted insisted that none of the hunters besides himself and Kermit carry a gun. The hunters were so upset by this direction that Kermit immediately turned to the subject of rewards to cheer them up. He promised that in addition to their pay, they would receive double the market price of any animal the Roosevelts killed. Kermit also promised a pair of field glasses to anyone who showed them a panda or a takin, the elusive goatlike animal with "the horns of a wildebeest, the nose of a moose, and the body of a bison."

There was an innate hierarchy to the expedition, and it tightly governed who was allowed to participate in science, how much they were paid, and their ability to hunt. The Roosevelt brothers perched on the top, with Suydam next, who was allowed to shoot panda. They were followed by Jack, who collected and prepared specimens, except for panda. Next were the guides who led the expedition and who were occasionally allowed to hunt, but not prepare specimens. Operating below them were the local hunters, allowed only to hike through the forest and banned from even carrying their weapons.

The complex group dynamics mirrored those of the golden snub-nosed monkeys, where the presence and role of males is tightly controlled. Large groups of females live together with single breeding males and their offspring. They form massive troops, up to eighty monkeys cohabiting in the tree canopy, but not everyone gets to be part of the family.

Before the age of three, the male offspring are pushed out of the troop. Like the Lost Boys of Peter Pan, they are wild and rebellious, living alone without parents. They band together, depending on one another for survival even as they tussle and fight to find their place within the new hierarchy. They feed, forage, and rest together, all while sharpening their fighting skills. To return to the troop, they must challenge one of the alpha males to a fight. If they fail, they might be forced to live alone, barred from both societies.

The dramatic social hierarchy of the golden monkeys matches the extreme nature of their habitat. The animals occupy some of the highest mountains on the planet, in a deciduous forest containing more tree species than anywhere else on earth—the perfect environment for elusive animals to hide. Although on this day, the Kingdom of the Golden Monkey was proving to be perilous.

As Ted and Kermit trudged back to the trail, wondering who had fired the shots, they came upon three hunters of their party, guns slung over their shoulders. At their feet was a dead golden snub-nosed monkey; in their arms a tiny newborn.

The infant monkey's face was the same blue hue as its parents', but its coat was lighter and thinner, a downy white. Spring was the birthing season for the monkeys and this delicate creature was possibly only a few hours old. The baby was terrified and cried out in fright. Kermit picked it up and the newborn's fingers clung to his shirt tightly, as it had so recently wrapped its small fingers around the silky golden fur of its mother.

Kermit did not often yell—he was known for his calm demeanor even in the face of stress and anger—but now he exploded in fury. He lectured the hunters about their shooting, reminding them that it had been expressly forbidden on this expedition. As bewildering as the tirade was—after all, the party was here to shoot monkeys—everyone understood why Kermit was upset. The infant shivered with fear as it clung to him, and every shudder seemed to fuel the Roosevelt brother's ire. Absent from Kermit's outrage was the stark fact that it could easily have been the Roosevelts who had shot a mother with a baby.

They brought the little monkey back to camp, wrapped it in blankets, and tried to feed it. The newborn let out anguished cries for its mother, piercing enough to break the heart of even the most hardened game hunter. Unfortunately, it was now time to skin the monkeys, which needed to be done if they were to be mounted at the museum. It was tedious work, requiring the utmost care so that the skin would only be torn in a few defined sections and the animal preserved properly for scientific study back in Chicago. Ted was not as skilled as Kermit in skinning animals, so he cradled the newborn monkey while his brother began the bloody task.

When Kermit was done, the hunters cooked the monkey meat and ate it, but the Roosevelts chose to have rice and beans instead. Their guides also abstained, and the group ate in uncomfortable silence, punctuated only by the sad, desperate cries of the infant monkey. When the meal was over, Ted, Kermit, the hunters, and one of their guides, a man named Hsuen who also acted as interpreter, sat around the campfire and discussed plans for the following day. Ted and Kermit wanted to make sure that their intentions were clear. There were to be no hunting dogs and no guns used by any of the local hunters. The small monkey, whose cries were becoming softer but more desperate, offered proof of the day's failures.

After finding the golden monkeys, a sliver of hope began to blossom that this jungle might contain other rare species. As Ted and Kermit slipped into their bedrolls, they talked about what would happen if they saw a panda. They had been taking turns at which brother fired first during their hunts so that each of them had a chance at the glory of having hunted the animals successfully. It was a practice that their father had taught them and was one of the reasons why he had had such a lofty reputation among hunters. With the panda, however, they made a different plan. They decided to shoot together. "We wanted to be full partners in the first panda," explained Ted, "should the gods permit." The Roosevelt brothers then drifted off to sleep, dreaming of black-and-white bears.

Dawn broke with a sign that the day ahead would not go well. "The little monkey," wrote Kermit in his journal, "we tried to keep alive, but it was too young." A gloom descended on the explorers. They had desperately hoped that the infant would beat the odds and survive. Instead, its small, cold body lay next to its mother and other members of its troop, never to return to the treetops. They skinned the tiny creature for the museum, but its death hung heavily round their heads, The weight of their guilt added to the heft of their packs as they set off for the jungle that stretched deep into the mountains rising from the valley.

Within a few miles, the hiking became nearly impossible. "At once we found ourselves in the densest jungle I have ever known," wrote Ted. "It was of bamboo six to eight feet tall, interspersed with Hemlock and Beech.... We had to climb up on hands and feet.... The dust from the dried bamboo leaves got into our lungs and eyes. The stems and matted vines through which we had to force our way tripped and clung to us like the tentacles of an octopus. The sweat ran in streams and caked the dust. It was impossible to see twenty feet."

Ted and Kermit kept going, but the small trail had vanished into nothing. They were plunging deeper into the forest, and despite the presence of their guides and the local hunters, they were rapidly becoming disoriented. Every direction looked identical with its massive stalks of green bamboo. Occasionally, a hunter would shimmy up to the top of the bamboo, but all he spied was an endless sea of green.

By noon, the sun began to beat down on the explorers. They had been hiking for six hours without a break and with no water. Ted's and Kermit's throats burned with thirst. They scooped up handfuls of snow, but it offered little relief, the crystals rough and dry as they swallowed. Desperate to find a stream somewhere, they scanned the forest floor for any hint of moisture. As they searched, an area where the bamboo had been flattened caught their eyes. A distinctive pile of scat sat in the middle of the opening: green, round, and large, each piece of feces the size of a dollar bill. The moment the Roosevelt brothers saw it, they knew what animal had left it there. It had to be a panda.

Thirst, fatigue, and the heat were instantly forgotten. They were on the trail of a panda now, and the Roosevelts knew that this might be their only opportunity to get the elusive bear for the Field Museum. The scat was old, unfortunately, but it was the first evidence they'd seen that the panda was real and lived in this habitat. Ted immediately asked the local hunters to fan out instead of hiking in single file, in order to cover a greater area, and possibly frighten an animal into movement.

Kermit's eyes darted everywhere as they hiked up the mountainside. He was searching for animal tracks and was certain that if they looked carefully enough, they would find more evidence of the panda in the jungle. Hours passed, but there was no sign. Eventually a grassy slope appeared above them. It looked like a clearing in the jungle ahead and the brothers became enthusiastic about hiking to it, believing it was an ideal spot for animals to roam in the open. The local hunters, however, were skeptical. "It's time to head back to camp," they protested. They had been hiking for ten hours and the meadow was still an hour's march ahead, over a rocky and icy trail.

Kermit acquiesced, but Ted held firm. He was certain that the meadow above them held their best chance for success and, despite the warnings of the hunters who said they would find nothing, was desperate to go. It was rare for them to separate, but Kermit was more inclined to listen to the locals, so he headed back with a group of them, while Ted continued with one hunter and a guide. The trail was every bit as bad as the hunters had warned. The threesome climbed up the dry bed of a stream, "coated with ice and choked with boulders." The light was rapidly failing, and dusk was settling on the valley. No matter what, Ted knew they'd be hiking in the dark, not a pleasant prospect given their past experiences.

In the low light, the trail became even more treacherous. "Every step I slipped," Ted recalled. "Every three steps I fell." Finally they reached the meadow. The last dying light of the sun had turned the grass golden and cast long shadows that clung to the rocks. Ted searched everywhere, but the hunters had been right. There was no sign of any animal up in the high mountain meadow, not even a track of one. They started back for camp, slipping and falling most of the way down the mountainside. Ted's body was sore and tired, but his mind was even more disturbed. He had been tantalizingly close, but now he was left with just frustration and the disappointment of returning to camp empty-handed.

It was a long march in the dark before they finally spotted the tents. "The moon was up when I saw the campfire twinkling," Ted wrote. "Of all welcome sights I have seen, none was ever more welcome. Altogether it was the hardest day's hunting I had ever spent." With waves of exhaustion and pain washing over their bodies, Ted and Kermit whispered to each other in the dark. "It's impossible," Kermit said. Ted could only nod in agreement.

They decided to listen to the local hunters and use dogs. It wasn't a tactic they liked; setting a dog toward their game would terrorize the animal and make it impossible to study the panda in its natural habitat.

But they had become desperate. They both knew that this jungle held their best chance of getting a panda and they couldn't leave without giving it their all.

The next day brought another long hike through dense jungle. The hunting dogs raced ahead of them, frequently becoming entangled in long, clinging vines that wrapped around their paws and then yelping to be freed. At first the march seemed promising. The dogs quickly led them to more panda scat. It was old but there was plenty of it. They marched through the bamboo with brisk, light steps, their rifles slung over their shoulders to be at the ready. There were definitely pandas somewhere in this jungle, if only they could find them.

They climbed higher on the hillsides, following the dogs. A howl moaned ahead of them on the path and the Roosevelt brothers became excited. Perhaps the dogs had picked up the trail of a panda. Ted and Kermit hurried through the maze of green, toiling through an ancient bamboo forest with trunks so massive they resembled marble columns more than they did trees. The mature bamboo was speckled with a paintbox of fungi, dappling the plants in shades of white and yellow. Here in the older bamboo forest it was somewhat easier to see where they were going, as the trunks grew in tall, straight lines before fanning out in fresh green leaves at the top.

To Ted and Kermit, the bamboo looked like trees, growing tall and straight, but if Herbert had been there, he could have corrected them: it is a grass that is so hardy and successful that its 1,700 species are native to every continent but Europe and Antarctica. Bamboo is a perennial flowering plant that some biologists have compared to a ticking time bomb. The green stalks grow for decades, sometimes more than a century, in a vegetative state. Bamboo seems timeless, as steady as any tree in the forest—and then, sparked by an unknown source, it will suddenly burst into delicate clumps of hanging flowers and die.

Millions of years ago giant pandas were omnivores, consuming both plants and animals. Then their bodies evolved. They lost the umami taste receptors that allowed them to appreciate the savory flavor of meat.

Then their jawbones and teeth shifted and flattened, making it easier to crush thick stalks of bamboo. Finally, an extra digit emerged in the wrist, a kind of "pseudo thumb" ideal for grasping at plants. Bamboo remained steady, unchanged, but the panda altered its body radically to consume it. It's a plant that few other predators desire due to its low nutritional value, but pandas eat thirty to forty pounds of it a day. The Roosevelt brothers could not yet know how intimately linked the two species were, but as they hiked through the bamboo jungle the first hints were beginning to emerge.

When Ted and Kermit reached the howling dogs, they realized that the animals were just play-wrestling with each other. They hadn't found a panda at all. Theirs were yelps of joyous play in the openings of the old-growth forest. Kermit walked away, annoyed. He was certain that the dogs were scaring away any large mammals hereabouts with their hullabaloo, but he decided to search the forest clearing anyway.

"Ted," Kermit suddenly yelled with rising excitement, "come over here." The two knelt at the base of a wide tree trunk. Claw marks were etched deep into the green surface of the wood. Kermit rubbed his hands along the gouges that marred the smooth surface of the tree. His fingers caught something. He held it up for Ted to see. It was coarse white hair.

Meanwhile, Suydam was hiking in a similarly vast bamboo forest to the south. The steep slopes covered in bamboo were beginning to take their toll. Days of trekking without result had left him frustrated and exhausted. "The crackling sound of our progress was enough to frighten any animal away," he wrote, irritated at his own footsteps. He hated the way his boots slipped in the soft dirt, how steep the hillsides were, and how easy it was to become lost in the sea of green. Yet these were minor annoyances beside the real problem: he had yet to hunt a single animal in these woods. "From dawn to dusk I scoured the country," Suydam reported, "and returned to camp empty-handed." He felt useless, and as the days passed, the sentiment grew until it became a monster within him, draining all hope for the expedition.

At the Roosevelts' camp, the outlook was just as bleak. The white hair Kermit found had temporarily buoyed their spirits but had been followed by nothing. Not even a trail existed of an animal's movements through the jungle. Day after day they searched and repeatedly came up empty. At camp, there was concern among the local hunters that the gods might be angry at the explorers. The death of the newborn monkey haunted them all, a painful episode that was enough to anger any deity. It did not seem like coincidence that their first foray into the jungle had been successful, but every hunt since that fateful day had ended in failure.

The hunters decided to build a small, temporary Taoist altar. A statue of Zhao Gongming, a deity riding bareback on a tiger, was placed on an elevated board. Beneath it, cups of water, tea, and rice were lined up, and candles were lit. Then a sacrifice was made—a live hen killed to appease the gods. When the prayers were finished, the chicken was eaten, and everyone hoped the next day would finally bring them the prize they sought.

They set off before dawn with misty, cold air swirling around them. Ted had decided that if they could not find a panda, perhaps they could at least get a takin today. The elusive species of goat was only found in this section of the Himalayas and he knew the museum would love to have a specimen. His hope did not last long. "The day was merely a repetition of the days before," Ted wrote. "Never have I seen a stiffer jungle." By afternoon his pants were in tatters from the thick underbrush that continually pulled at the fabric, and his spirit was as ragged as his clothes.

In a grassy clearing, the Roosevelts decided to unfurl their bedrolls instead of returning to camp that night. They reasoned that the early morning might bring animals to the clearing, and if they spent the night there, they would be waiting for them, perfectly quiet, instead of stamping around the bamboo. For Ted and Kermit, it felt like a last-ditch act. "If the game ever did come into the clearings to feed, that would be the time," Ted wrote. "If they did not, it would be pretty conclusive proof that game rarely if ever left the jungles."

There were no tents so Ted, Kermit, and one of the hunters slept under the stars. The air was crisp and cold, and clouds moved across the sky, obscuring the beauty of the heavens. There was little talk between the brothers; they were too tired. Instead, they "slept the heavy sleep of exhaustion," their tired muscles pressing into the hard ground beneath them. When dawn broke, they took out their binoculars and scanned the long grasses of the meadow. The still air brought all the peace and quiet a person could wish for, but not a single animal.

"Each day was a repetition of the former," wrote Kermit of the next few days, "ceaseless struggling through dense jungle, without hearing or seeing anything." There was no more talk of appeasing the gods or sacrificing hens. It was clear that the Roosevelts were beyond help. Desperate, Ted and Kermit took the local hunters aside and begged them to speak honestly. "Is there any other place where we might have a fighting chance?" Ted asked. They all shook their heads.

After some conversation, Ted and Kermit learned that only three men among the thirteen professional hunters had ever shot at a panda. The other ten men had never even seen one. No one in the group knew of a person who had successfully hunted a panda. It seemed to Ted that mere luck would determine whether they found the bear or spent years searching for one.

The Roosevelt brothers sat beside the campfire without seeing the flames. They could speak of nothing but the panda—their minds had become obsessed—but every word was a reminder of their own failure. They hypothesized the animal's behavior, debating whether it hibernated, what part of the bamboo stalk it ate, and why they had found white hair lodged in the claw marks. Perhaps it was "the bear's habit of measuring itself against a tree and scoring the bark with its claws," hypothesized Ted.

To Ted and Kermit, it seemed likely that the panda merely wanted to scratch its back on the bamboo, a behavior typical of the brown and black bears they were familiar with. But panda bears, unlike those others, *communicate* through bamboo. The bears are solitary creatures that spend most of their lives alone, the exception being the mothers and their cubs who can spend up to three years together. Except for this period of time, the independent animals have little means to share information with one another. Their growls are soft, their tails short, and even their faces are round and inexpressive.

To "speak" with one another, the panda rubs secretions from its anal gland onto the bamboo, tree trunks, and even rocks on the ground. The secretions contain a wide range of compounds, including aldehydes, alcohols, ketones, esters, sterols, acids, proteins, and aromatics that serve in chemical communication. This scent marking lets the creatures get to know one another, even if they never meet. They can discern age, gender, health status, and diet merely from a sniff. Scent marking can serve as a warning, effectively saying *this territory is mine*, or as an invitation. Females use scent marking to let male pandas know they are ready to mate. A female is far more likely to mate with a panda she's "met" through his smell in the jungle, even though they have never seen each other face-to-face before.

If only Ted and Kermit had had the superior nose of a bear, they might have been able to read the scratches and scents they had found in the forest. As it was, they were more confused than ever. Ted questioned whether the beast they'd been pursuing was a bear at all. He believed that since the animal didn't hibernate, it must be radically different. They were grasping at air, and this most mystical bear was far out of their reach. "So much for the panda," wrote Kermit. "It was with regret we turned our backs on it. May the fates give us the chance to match our wits against it again before the windows be darkened."

CHAPTER 10

TEMPLE OF HELL

The streets of Yachow ran red with the blood of innocents. Jack Young watched in terror as a Chinese official held up a gun to a young boy's head and fired. The boy fell to the ground. The officer raised his hand twice more, and when he was done three boys, the youngest just sixteen years old, lay dead in the street. Their bodies remained untouched, a warning to all who passed that the Kuomintang, the Chinese nationalist party, did not treat rule breakers lightly. It was a public execution and Jack had happened to come across it while walking around the city.

Arriving five days late in Yachow, Jack had been relieved to learn that despite being delayed by bandits, he was still ahead of the Roosevelt brothers. Now his feelings of relief were turning into terror. He had never seen anyone die. He asked a man next to him what crimes the young boys had committed. "They were making speeches," the stranger replied. The National Revolutionary Army had been demanding money from the townsfolk, claiming that they were collecting taxes in advance. The boys had rallied against the injustice and started making public speeches against the practice. While their talks had been peaceful, the penalty for making them was brutal.

TEMPLE OF HELL

Ancient China had become known for the "five punishments," employed during the early Han dynasty, which consisted of tattooing, cutting off the nose, amputation of one or both feet, castration, and death, with variations over the centuries. According to Chinese folklore, the practice ended only when a young woman named Chunyu Tiying wrote a letter to the emperor in 174 BC pleading for the release of her father, a physician who she claimed had been falsely accused of malpractice. Her writing was so eloquent that the emperor agreed to change the laws, substituting the less severe punishments of flogging and hard labor for the five traditionally dispensed. While the penalties would shift over the centuries thereafter, an underlying cruelty continued, with harsh physical punishments routinely meted out for nonviolent crimes.

"It's a sad commentary on conditions," wrote Ted of the executions after Jack described them to him, "three high-school boys, ranging from sixteen to nineteen, were shot." As shocked as Jack and the Roosevelts were by the executions, the United States could not claim superiority. Its citizens were still subject to brutal and inhumane lynching, used to intimidate and control Black communities. In addition to lynching, which occurred ostensibly outside the law, the death penalty was experiencing a resurgence in the 1920s. Crime rates had increased during Prohibition and a push for the death penalty by criminologists and politicians ensued, including public executions. Between 1920 and 1930, more Americans were being killed by the death penalty than in any other period in the country's history, with an average of 167 executions carried out each year, two-thirds of whom were Black citizens.

Whether in the West or the East, the death penalty was deployed as an instrument of dominance over racial and ethnic minority groups. In China, executions were used to frighten and dominate marginalized people, such as the Tibetans and Nashi, who lived in regions where the nationalist government was striving to gain control. By executing young boys who spoke out against the Kuomintang, the country was attempting to squash future uprisings in a remote part of the country where nationalist power was still tenuous. As Jack watched the young men die, he realized with a shudder that he was barely older than they were.

Group of Nashi villagers. Photograph by Suydam Cutting, 1929.

Fifty miles west, in the forests outside of Muping, the Roosevelt brothers met up as planned with Suydam, who had also been unsuccessful hunting panda but had shot a large bird of prey. He was eagerly showing off the creature, and it was spectacular. The bird was a bearded vulture, stretching four feet tall, with dark, tawny feathers, black on its wings and back and rust-colored on its belly and head, with a black band extending across the eyes. It differed from other vulture species they'd seen, with a head covered in feathers instead of being bald, and a massive wingspan stretching some nine feet. Most impressive of all was what no one could see: its stomach contents, which contained concentrated, pure acid capable of dissolving even large bones in as little as a day.

Vultures feed on death, picking over the carcasses of deceased animals. Bearded vultures are unusual in that they don't eat the meat, but instead feast on the bones, particularly the bone marrow. The birds will pick up pieces of the skeleton and then fly high into the air, dropping the bones over rocks to break them apart. They'll do so repeatedly, until they break open the accumulation of fat cells that make up the bone marrow. They are the only bird species that specializes in bones and are adept at cleaning up the refuse of other species.

The Roosevelt brothers fed on death in a different way. They packed up their own collection of bones, feathers, and fur, and began hiking with Suydam toward Yachow to meet Jack. It had been only eight days since they'd last seen Suydam but it felt as if the world had changed completely. "When we trekked up Spring was hesitating on the threshold," Kermit wrote. "Now it had flooded in." Butterflies flittered over the colorful heads of flowers that stretched across the trail while bees collected pollen on their hairy hind legs. As the explorers walked, the number of bees began to increase, and it suddenly felt as if they were surrounded by hundreds of them, buzzing in every direction. As they approached a nearby town, the drone of bees filled their ears, growing louder and louder.

The sound seemed to be coming from a nearby Taoist temple, so the brothers entered cautiously, passing dozens of bees buzzing lazily from the open front door. Their eyes adjusted slowly to the dark, cavernous space, and they came to realize they were surrounded by clay statues. They stepped forward and discovered the buzzing was coming from inside the statues. It was so loud that it echoed across the vaulted ceiling and the dramatic, peaked roof.

The bees had made holes in the soft clay of the statues and the explorers watched the insects crawl across the figures as they left their hives. Ted spied the golden gleam of honeycomb oozing from one of the openings. He realized the place must be filled with hundreds of pounds of honey; it was dripping down the statues and leaving sticky trails across the floor. Ted looked down at his boots, which made smacking sounds as he moved across the foyer. Then he stood back for a moment and looked more closely at the statue in front of him. His jaw slackened in shock.

Every statue and painting depicted torment. The people, molded from clay, grabbed at their faces, yelling in wide, silent screams, or fell on their knees, suffering in misery and shame. Many were being tortured by demons, with their tongues in the process of being cut off, their bodies burned, or their bellies disemboweled. Some of the statues had been painted in vivid colors, making the scenes even more realistic and disquieting. As the travelers inspected the twisted faces of the figures, they realized that each shrine depicted a mortal sin, from infidelity to thievery to drug use. Amid the horror, cracked pieces of the statues littered the ground. Ted noticed an arm dangling precariously from a shoulder and then a leg, twisted off its base, broken on the floor.

"What's this place called?" Ted asked. His spine tingled with unease and, perhaps for the first time, there was nothing inside this temple he wanted to buy.

"Temple of Hell," their guide Hsuen replied.

"And what's that?" he inquired, pointing at a thick, sticky substance that had been smeared across the mouths of the statues.

"Opium," Hsuen replied. "They give the gods their share."

"Why are the statues broken?" Ted asked as he moved through the temple delicately, as if scared to touch anything. Hsuen explained that if their prayers went unanswered, people would attack the statues angrily.

It was quiet inside the temple as the explorers made their way from room to room. Surrounded by examples of divine retribution, it was impossible not to feel a sense of foreboding. Ted and Kermit wondered if the local hunters had been right, and their expedition was cursed. After all, the death of the newborn monkey screamed for reckoning. If so, the temple offered no hint of redemption. The statues that depicted their possible future punishment were so terrifying that Ted and Kermit stared transfixed. They could only hope that the fate of their journey was not portrayed somewhere within these walls. It was some time before they could shake off the dark prospects before them and hit the trail again.

The country they were traversing revealed the sins of battle. Ted noted that "bitter fighting" had recently taken place between the National Revolutionary Army and the local militias, and no one could be certain who currently held power in the region. Placards warning of banditry dotted the trail amid the flowers and trees. When they heard shots fired behind them, Ted, Kermit, Suydam, and their guides took off running with their backpacks on. Without mules to lead, the explorers were able to move as fast as their feet could carry them, and they raced down the road. It was harder for the guides whose backpacks were the heaviest and who were additionally weighed down with animal specimens from the recent hunting expedition. "It was remarkable," Kermit wrote, "to see the way they ran over the rough ground carrying a seventy-pound pack."

As they approached Yachow, the fear of being caught in civil war eased and they could appreciate the beauty of the country again. "We had changed from a rough wild country to one of the earliest settled parts of China," explained Kermit. "The country was just what I, as a boy, had always imagined China. Valley and hillside were cultivated. The greens of the various crops formed a patchwork of color. Enormous wooden waterwheels for irrigation were slowly turning in the river. Fruit trees were masses of blossoms. There were delicate mauve lilies blooming by the roadside. Every little while we passed memorial arches or tablets beautifully carved from red sandstone. Even in the tiny farming villages there were often delicately carved stone ornaments on either side of the doors. To add a final touch, we passed a number of graceful white pagodas silhouetted against the sky."

The region the explorers were moving through was a cradle for the ancient tea trees of China. The tea gardens were more than a thousand years old, the origin of all the world's tea leaves. It had begun with seven tea trees planted in 53 BC whose leaves were harvested every spring and fall. The gardens had been carefully cultivated over a millennium and were bursting with tea varieties, flavors, and aromas. Some tea leaves were considered so precious that they were preserved for sacrifice to the gods alone; not even the emperor was allowed to taste their brew.

The walled city of Yachow loomed ahead. The air was humid and warm as the explorers passed through a heavily adorned gate, shook off the dust of the trail, and entered the broad, paved streets of the city. They were heading for a large red-brick building. As usual, they were staying with missionaries, this time the Baptists, who also ran a hospital out of the mission. Their religious operation was slightly different from others the explorers had observed. Many missionaries liked to be called "Doctor," although they held no degree. "The Holy Rollers," as the botanist Joseph Rock had described the missionaries in this part of China, "were without any education, couldn't compose a letter, but the Lord had called them from their alma maters, the dumpcarts, to convert the heathen."

By contrast, Dr. Crook, who led the mission, actually had a medical degree. He used the hospital to proselytize but also to train young Chinese doctors. His plan was to develop a staff of medical professionals who could travel to remote regions of China in desperate need of medical help. It wasn't working out as he'd planned, however. "As soon as a man was properly trained," wrote Suydam, "the local magistrate would impress him into the army."

They had only just arrived and said their greetings when Dr. Crook announced with a smile, "Telegrams." It was the first news they'd had of their families since leaving home four months ago, and Ted, Kermit, and Suydam eagerly grasped the papers. The excitement soon turned to bewilderment. "State Chenry Stimson Treasury melmon," read a telegram to Ted, "was assd justice William Mitchell mines sota postmaster Welter broa navy Charles francis Adsmo Massachus Ctts interior ray Wilbur agriculture Cyde Missouri Lalar davis Commecr nnawnon Ced."

He stared at the letters in confusion. He could pick out only parts of words and names; the rest was a jumble. Ted was used to reading the clipped phrases of telegrams, which were always shortened to save money, but this was beyond his skill. Translating the words from English to Mandarin to English, necessary for the different telegraph machines used in the United States and China, had rendered them unintelligible. Disappointed, he put aside the telegram and turned his attention to a bath instead.

Once clean and fed, Ted was in for a treat. "Yachow proved the best and largest bazaar," he wrote of the town. As usual, the shops called to him, and he wandered through the market, happily inspecting the items for sale and talking to everyone. While Ted bargained over a handsome ivory chess set, Kermit watched the opium distillers at work in their shops. They were stirring a mixture of dried opium pods and water in large oil drums over a fire, cooking down the sludge and then adding chemicals to the mixture. Huge clouds of steam emanated from their pots. It was the distillers, Kermit learned, who became the most devoted addicts of the drug. The smell was so strong that he only asked a few questions of one of the guides before hurrying off. It was much more pleasant to watch the men playing chess in the square.

Kermit thought he had seen the vast toll of opium use already. It had been present in nearly every village they passed, and even among those poor, addicted guides hiking the trails around them. Still, he was unprepared for the scale of the devastation in Yachow. The explorers took a tour of the hospital, with Dr. Crook leading them through the whitewashed halls. It seemed that every room was filled with addicts desperate for a cure. Delirious screams pierced the steady hum of activity in the hospital wards, and when Kermit asked what percentage of the patients were able to be cured of their addiction, the physicians only shook their heads and would not even hazard a guess. The success stories were so few that they hesitated to talk of them.

Back at the mission, the explorers began sorting through their backpacks and readying for the next leg of their journey. They went through their books, leaving behind tattered and stained copies of *Pride and Prejudice, Sense and Sensibility,* and *Jane Eyre*, all of which had provided solace and comfort in the peaks of the Himalayas. Then they organized their clothes and belongings for the weeks ahead. They were embarking on the most dangerous segment of the expedition, a decision they had made only in desperation, and they needed to prepare themselves both physically and mentally.

The missionaries warned the Roosevelts that what they were attempting would likely get them killed. "Lolos are a primitive tribe," explained Dr. Crook, calling them "savages" who "capture and enslave their prisoners." It was a familiar warning for the Roosevelts. In addition to the missionaries, several other naturalists had told Ted and Kermit not to enter the lawless territory.

The area was governed by the Yi people, one of China's largest ethnic minority groups. They had many names, depending on who was speaking of them, including the Lolo, Sani, Puwa, and Ni. They had lived in the remote mountains of Southwest China for millennia, since before the Tang dynasty, maintaining their own autonomous rulers and royal family and speaking their own language. Unlike many other minority groups, they had proven successful in repulsing Chinese rule. Perhaps it was the terrain, which was rugged and had few roads, most of which were considered impossible to navigate. Or maybe it was the rumored warlike nature of its people, a reputation built through battle against their neighbors for centuries. Whatever the reason for their fierce independence, it was clear that the Yi people were poorly understood. Their society was insular, and although the Roosevelt brothers asked many questions, it seemed few people had answers.

During their travels, the explorers had noted that the Yi people were sometimes held as slaves by the Chinese. They were seen as subhuman, worthless beings whose spirit of independence had to be beaten out of them. Sadly, slavery was part of their own autonomous state, and Yi society was ruled by a strict caste system. At the top of the hierarchy were the black bone Yi, the aristocrats, whose status was unquestioned and who owned the vast majority of wealth and property. Below them were the white bone Yi, considered the commoner population. The castes were not allowed to mix and lived in separate villages with marriage between the two forbidden. At the bottom of the hierarchy were the Jianu, who held no rights and were kept as slaves by the upper castes. It was this group that commonly served as hunters in the region. Unlike in many of the ethnic minorities the Roosevelts had become accustomed to in Tibet and China, women had few rights and did not serve as hunters or guides.

The Roosevelts had decided to take a gamble. They reasoned that even if the Yi people were fiercely anti-Chinese, that did not mean they were hostile to all foreigners. The explorers had traveled through multiple autonomous regions inhabited by a wide range of Chinese and Tibetan minority groups to get this far and had found that, despite warnings, they were always kindly received. Several times they had been warned about the lamas, but it was those spiritual leaders who had rescued them when they were lost and without food in the mountains. They hoped that their own neutrality would give them an advantage among the Yi, but of course, they could not be sure where the next few months would lead them.

What they knew for certain was that this was their last chance. If they didn't find the panda here, in what many Westerners called the "land of the Lolos," they were likely to never find the animal at all. It was in this isolated region that the French missionary Pere David had obtained the first panda skin, now sitting in a dusty back room of the Muséum National d'Histoire Naturelle in Paris. This was where the search for the panda had kicked off sixty years earlier, and Ted and Kermit hoped that it would end in the same place.

It was the last day of March when the explorers hit the trail once more. They had spent only two days in Yachow; there was no time for a longer visit. The monsoon season was coming, and the window for hunting was growing ever shorter. At least they had mules again. The pack animals lightened the loads on their backs, leaving the explorers to fly down the rocky dirt trails feeling light and free, a contrast to their months in the Himalayas. One of their favorite guides, Hsuen, had taken advantage of the mule train and had strapped a birdcage to one of the packs. He proudly told Kermit that he had bought the birds because "they sang so pleasantly in the morning," and like a living alarm clock, "would awaken us with their music." Ted and Kermit leaned closer to the cage, but the birds were silent. The brothers looked at Hsuen inquiringly, and the guide, feeling the pressure for the animals to perform, rattled the cage gently. As if on cue, the birds burst into song.

The birds turned quiet as rain began to fall. It started as a soft tapping, gentle on their heads, then crescendoed into a downpour, until the hikers' bodies and packs were soaked. For four and a half hours they trudged through the mud. The air was so thick with mist that they could see only a few yards ahead of them on the trail. They were climbing a ten-thousand-foot mountain pass, and snow was beginning to accumulate. It was not a welcome sight. They'd had enough inclement weather already on this expedition and had hoped they were leaving winter behind. They slogged on, sharing the narrow path with a throng of tea porters, all carrying the precious Yachow product from its cradle to the vast population of China and Tibet who craved the mellow, sweet leaf of green tea.

Among the porters toting hundreds of pounds of tea were Chinese magistrates traveling in lavish palanquins, a sort of covered box designed to hold one person and their belongings, carried on large poles by four porters. These were grand affairs, resembling massive, gilded birdcages, and called attention to the importance of those who rode inside.

The Roosevelt brothers were rounding a steep curve of the trail when they noticed two palanquins lying motionless on the ground. The conveyances had been abandoned by their porters and Ted heard cries of frustration coming from behind the folds of silky curtains that hung around the seat. Concerned, the guides spoke with the man inside and learned that he had been attacked by bandits. The marauders had easily overcome the porters, whose hands had been busy holding up the palanquins. In a burst of violence, they had thrown one seat to the ground, seized the belongings inside, and then dragged the man sitting inside off into the mountains. This lone traveler was now abandoned. In a matter of moments, he had gone from an object of envy to a creature of pity. It was a reminder of how quickly the trail could take everything from them.

The brutality of the attack shook them, but they had little to offer besides advice. They certainly weren't about to hoist a palanquin in the air, no matter how much the man was crying. They encouraged him to begin hiking, reminding him that there was a town nearby, and then hit the trail themselves, their eyes scanning the woods for thieves as they walked. Soaking wet and stumbling with exhaustion, they finally made their way to a small inn.

As the guides unpacked the mules, Hsuen held up the cage holding the two songbirds he had purchased in Yachow that morning. His face was twisted with grief as he showed Ted and Kermit the corpses of the two delicate animals. "The cold and sleet had proven too much for the songsters," wrote Kermit. Hsuen began to sob quietly. He had spent a good portion of his wages on the animals and had hoped they would accompany the expedition the rest of the way, spreading cheer with their song.

As the guides unpacked the mules, they began splitting up the provisions. The expedition was dividing once again. The "land of the Lolos" was going to be treacherous, and the Roosevelts had decided that there was no reason for the entire caravan to be put in danger. Suydam and Jack would take a shortcut back to a main thoroughfare on their way to Yunnanfu, where they would meet Ted and Kermit a month later. The route would lead them through some hilly terrain that the Roosevelt brothers hoped would offer Suydam a chance at hunting big game while Jack could collect the smaller animals. With a smaller group, and on a trail too narrow to permit mules, Ted and Kermit would then head on to the land of last resort.

For the explorers, it felt like the end. In the five months of their expedition, the party had collected five thousand bird skins, two thousand small mammals, and forty big mammals, including the sambar, the serow, the blue sheep, and the golden monkeys. They had identified nineteen new species, still unnamed. So little was known about these creatures that they had become legendary among Westerners, and Ted and Kermit knew that scientists would be anxious to probe their secrets, after which the museum would proudly exhibit them. Still, the group was depressed. "It appeared to be a hopeless quest," wrote Suydam. "The giant panda expedition was ending without a sight of the giant panda."

"I'll be glad to get rid of this," Suydam remarked as he shook out his bedroll. The fabric smelled like rancid yak butter from their time among the Tibetans in the Himalayas.

Hsuen, many miles from home and in need of comfort without his precious birds, gladly took it.

Sadly, Hsuen's birds might have endured if they could have held on for one more day. The trail, which had reached snowy peaks the previous afternoon, now dropped sharply. The air once again turned warm, and from the mist and rain, fields of flowers bloomed in all directions. The explorers found themselves in a meadow of poppies. The flowers "were a mass of color," wrote Kermit, "some solid white, others purple, others red, and still others a combination of pink and white. Stretching out on either side of the road and smiling in the sun, it was not easy to recall the sinister purpose hidden beneath their beauty." It was as if they had been transported to the Land of Oz, but instead of a wizard in an emerald city, they sought a mythical bear.

As they passed through fields of flowers, the trail sank even lower in elevation, dropping down into vast orchards of mulberry trees. The trees, *Morus alba,* were not nearly as lucrative a cash crop as their poppy neighbors, but their propagation held other commercial appeal. Mulberry leaves are the sole diet for silkworms, and at the right time of year, the munching sound of caterpillars echoed through the orchard. The feasting sounded not like the activity of insects, but instead like a library of books being shredded all at once.

The *Bombyx mori* has yielded silk fibers for thousands of years. According to legend, the empress Leizu was sitting under a mulberry tree when a moth cocoon dropped in her tea in 2700 BC. The strings of the cocoon came loose, and the empress unraveled them, wrapping the soft threads around her fingers. She then convinced her husband, the emperor, to plant her an orchard of mulberry trees for the silkworms, from which she wove the first silk fabric using a reel and loom of her own design.

It takes an astonishing three thousand cocoons to produce a single pound of silk. That means that thousands of silkworms are killed to produce one single, lustrous pillowcase. The "queen of fabrics" is highly valued, not just in China but throughout the world. The smooth material has shaped human history along the ancient trade routes that formed the Silk Road, connecting China to the rest of the globe. As silk was traded, so were ideas, customs, and beliefs, all thanks to a small, fragile insect that spins its own doom every time it attempts metamorphosis.

Humans can adore animals with a devastating ferocity. The relationship between silkworm moths and humans is interwoven so tightly that *Bombyx mori* can survive only in captivity. Because the creature is so highly prized and has been cultivated for so long, the silkworm is now completely dependent on its captors, its very genetic code twisted in response to its conditions. The adults have lost the ability to fly, and the cocoons can no longer dangle independently from the branches of the mulberry. In the wild, the moths have gone extinct. What is left is a pale shadow of what was once a vibrant species, altered forever by the unceasing consumption of humans.

Five hundred miles to the west, far from the silkworms, Herbert was arriving in the Kingdom of Muli. He was so far behind at this point that it was as if he were in a different country, surrounded by a wholly separate culture, landscape, and people. Herbert was beginning to worry about his slow pace. "My time was short," he wrote, "if I was ever to reach Tachienlu . . . before the rest of our party would leave." He had no idea that his fellow explorers were already long gone.

As he hiked through a meadow that led to Muli, he stopped to examine the butterflies flittering around him. Species of the Camberwell beauty, colored a deep red and fringed in bright white, and the Everest clouded yellow, a bright gold, flew lazily from flower to flower. Herbert wrote their scientific names and drew a quick sketch in his field notebook. It was behavior like this that had gotten him left behind in the first place, but Herbert was incapable of being in a hurry. His khaki shorts rubbed against the long grasses as he moved slowly and quietly through the meadow, admiring butterflies. Even in the afternoon, the air was crisp at these elevations, and so, ever the Englishman, he slipped into his tweed coat for warmth. He saw no species he didn't recognize, so instead of grabbing his net, he sat down on the ground and admired them peacefully. Whether collecting specimens or not, Herbert was determined to document everything, and he'd already filled six field notebooks with his narrow, crowded scrawl. Every animal and plant he came across was described in his pages, along with notes on their habitat.

Just like they had for the Roosevelts and their caravan, the lamas took in Herbert and his two guides. The lamas offered a comfortable place to stay as Herbert prepared to meet the king of Muli. The king had been traveling during Ted and Kermit's visit, so they had not met him, instead staying with his brother, the guardian of the eastern border, in the House of the Prince.

Herbert had heard no word from his fellow explorers in China, nor from anyone back home in India since the expedition had begun. On a brass tray he laid out the gifts that he hoped would be acceptable to the king: tinted snow spectacles, a monocular field glass, a brandy flask, and a tin of biscuits. The tray was whisked away, and in return a lama brought him dried pork, grains, and a brightly colored wool blanket that Herbert immediately adored and threw around his shoulders.

With the gifts formally exchanged, it was now time for Herbert to meet the king. The botanist Joseph Rock had described the ruler in his *National Geographic* articles, so Herbert was not shocked when he was led into the presence of a mountain of a man, standing six foot two with massive, muscled arms and legs that looked as if they had been sculpted from tree trunks. Rock's articles, however, had not prepared Herbert for the king's humility. There was no finery in the room and the king himself was dressed in modest attire. He sat on a worn sofa with one of his dogs, a Tibetan kyi apso, curled up nearby.

The lama ruler's name was Xiang Cicheng Zhaba and he had questions for Herbert. The area was so remote that it was only through travelers that residents could get information about the outside world. He asked Herbert about his journey and what he was looking for, but then his inquiries turned to science, a topic he was especially interested in: "Is it possible to see through people?" Herbert looked at his guide, who was translating their conversation, and then stared at the king in confusion. *What can he mean?* he thought. He darkly wondered if the king was trying to see into the minds of his people. For a moment Herbert's imagination ran wild as he considered whether this ruler was an evil mastermind, looking to manipulate people's thoughts.

Then it came to him. "X-rays!" Herbert shouted out. The king asked him to explain how they worked, and Herbert stumbled over his words, unsure himself. He was a biologist, after all. He knew little about how electromagnetic energy passes easily through soft tissues before it's absorbed by bones. "My explanation fell short of the mark desired," he wrote.

Hundreds of miles east, Ted and Kermit were also wishing for medical expertise. The explorers were staying in a small country inn where the wails of the sick and infirm were echoing off the walls. The plaintive cries of a baby cut through the cacophony of misery. Then there was a soft knock at the door.

It was Hsuen, their favorite guide, and Ted told him to come in. "What's going on out there?" the elder brother asked.

"It's a smallpox epidemic," Hsuen explained. "In the next room is a poor woman who has lost one child and whose baby is sick. Do you have medicine?"

The brothers looked at each other. Smallpox was a devastating and highly contagious virus. It had killed millions worldwide and was especially dangerous for young children. The disease starts with fever and vomiting before the patient is covered in painful fluid-filled bumps. Yet the virus was also preventable.

In 1796 in England, physician and scientist Edward Jenner had extracted matter from a cowpox sore on a milkmaid's hand and injected the gooey stuff in an eight-year-old boy, the son of his gardener. While a few other English physicians had vaccinated with cowpox previously, Jenner was willing to take the process a step further. Two months later he exposed the same child to smallpox. The inoculation was a success, and the child was the first person to receive Jenner's smallpox vaccine.

While Jenner is heralded as the father of immunology, China had been inoculating against smallpox as early as 200 BC. Physicians took the scabs from smallpox patients and ground them into a fine dust. They then gave their patients a small amount of the material intranasally, allowing rapid delivery across the mucosal tissues. Patients would usually still get sick, but the disease was milder and the mortality rate far lower than natural infection.

Despite China's long history with smallpox inoculation, and the invention of a tested and effective vaccine in England, most of the country's population, especially in rural areas, remained unvaccinated and thus vulnerable to the deadly disease. Ted and Kermit thankfully eyed the telltale, depressed circular scars on their upper arms. They'd received the vaccine back in the US and would not have been allowed to travel without it.

Now the Roosevelt brothers racked their brains, trying to think if there was anything they could do for the woeful pair next door. It wasn't just the baby crying anymore; the mother had joined in, and her deep moans of despair were heartbreaking. "There are times," wrote Ted, "when a traveler, called upon for help in a hopeless case, may give a harmless pill in the hopes that it may cause fictitious relief or comfort, but with a dying infant such a procedure seems only heartless."

The next morning, tired and groggy, the explorers laced their boots as they sipped steaming cups of green tea. The early morning silence, devoid of the cries of infants, seemed to confirm the worst. They turned to Hsuen, who was acting as lead guide this morning. The man was industrious, talking with every person he could find along the trail, and in every village they passed, constantly asking if anyone had seen *beishung*, or the white bear. Three hours into their hike, the Tibetan guide ran up to Ted and Kermit in breathless excitement.

"Last night," he explained, huffing as he caught up with his breath, "a man came to a house up ahead and asked to borrow a rifle." He was so excited he could barely get the words out, but finally the Roosevelts heard the whole tale. It had begun a month earlier, when a bear had entered a small, remote village in Lolo land and attacked a beekeeper's apiary. Now this same bear was said to be back, once again attacking the beehive, and the beekeeper was looking for something to scare the animal away. No one in the house he'd approached had a rifle, so the man had returned to his village empty-handed.

The bear, Hsuen said in one breathless gasp, "was a *beishung*."

Ted and Kermit looked at each other. This was the first hint in hundreds of miles that the bear they had spent the last five months searching for was real. The last time anyone had told Ted and Kermit they'd seen a panda was back at the Forge of Arrows, and that, of course, had led to nothing. This, too, would likely end in naught, but at least they had a lead to follow.

Although the beekeeper had left the night before, Hsuen had gotten the name of the village he lived in and rough directions to its location. The path leading to the town was narrow, and the farther they walked, the more constricted it became, until the explorers were firmly in the grip of the trail. The edges of the dirt path were bright green, teeming with life, and every step they took seemed to push them deeper into the thick jungle. There was no boundary on their map—the paper was flecked with only white in this region—and no sign or border apparent anywhere. Without knowing it, the Roosevelts had entered the "land of the Lolos."

CHAPTER 11

LAND OF THE YI

The game was called bear hunt. It began when Theodore Roosevelt hid in the family's vast Oyster Bay home on Long Island, squeezing his large frame into a closet or bathtub, and then remained as quiet as possible. Minutes later, his two eldest boys would come tearing through the rooms, looking for him. "Where's the bear?" they'd cry as they threw back curtains and crouched under beds. Every minute of looking built the anticipation, until they were so excited they could barely open a door without their hands shaking. When they finally found their father, Ted and Kermit would erupt in joy, screaming at the top of their lungs and pouncing on his back. He, in turn, would roar as loudly as possible, his glasses askew on his face, and then tackle his sons, tickling the boys' ribs with his fingertips until they all subsided in a fit of giggles on the floor.

Unlike many fathers in the early 1900s, President Roosevelt wasn't afraid to show his children affection or tell them that he loved them. As busy as the man was, he loved to play with his sons whenever he was home, to take them along on trips, and to write them long letters when he was gone, drawing pictures and signing each letter from "Your loving father." He was frequently hard on them, especially Kermit, who

LAND OF THE YI

he worried would grow up to be a weakling, but the Roosevelt brothers knew that when they hunted their father, the man they loved more than any other, all was right in the world.

More than thirty years had passed and their father was dead and gone, but Ted and Kermit were still playing bear hunt. This time, they didn't expect the game to end in kisses and tickles. News of a white bear had come from a village a two-day hike to the east, and they were now winding off on a side trail to track down the animal.

They trudged for hours down a narrow, overgrown trail until they arrived at a small town. It was about halfway to the village where the beekeeper with the bear problem lived. The town was nothing more than a handful of houses, with no caste system that the explorers could discern. It seemed like a good place to ask about the panda, so as a few villagers approached, one of the guides began to inquire about hunting in the area. The response was cold and wary. It was clear that the villagers were suspicious of their motives.

Ted pulled an axe from his belongings and presented it to the leader of the group with a smile while one of the guides made the necessary translation. The man's face remained stern, so Kermit withdrew an emerald-green homburg hat from their packs.

A group of Yi men. Photograph by Suydam Cutting, 1929.

The final gift seemed to do the trick. The leader of the group liked the look of it immediately and plopped it on his head, and within minutes they were all sitting together, sharing a drink, and discussing hunting and the wilderness around them.

Their informal chat took a sudden turn as soon as the word *beishung* entered the conversation. It became clear that the panda was not an animal they hunted. When the bear attacked their apiaries, they would try to scare it off, and even wound it if necessary, but they would never kill it. "Why?" Kermit asked, genuinely confused. The skin of a panda was rare and valuable, certain to bring a good price at market. He certainly knew from experience that it was impossible to buy. It was clear that this village was not wealthy; the hut they sat in, constructed from halved bamboo stalks and roofed with rounds of pine, leaked. Drops of rain dripped on his head uncomfortably as they spoke.

The bear is a "supernatural being, a sort of demi-god," one of the guides translated, but "it is not to be feared." To the local hunters, the panda was not just another animal in the woods. Its presence was rare; the members of this village hadn't seen one in months. But it was respected. Under no circumstances would the Yi people who lived in this part of China hunt the animal.

Ted and Kermit were silent for a few minutes as they assessed this new information. "We determined to approach the subject cautiously," wrote Kermit, "and to carefully avoid offending any native sensibilities." Ted and Kermit were quick to ascribe the practice of not hunting the bear to the "religious sensibilities" of "natives" instead of considering their own limited information about the panda's behavior and habitat. They still believed they were tracking a polar bear on the Asian continent, an animal that was elusive because it was aggressive and hostile to humans. They could not imagine the truth of what awaited them ahead in the woods.

The rain began to come down heavily, filling the forest with the dull cacophony of water against leaves and wood. Every drop meant another second gone for the explorers. It was the start of monsoon season, and the heavy rains would soon make hunting impossible. That night Ted and

LAND OF THE YI

Kermit nestled in the small hut that the villagers had generously offered them. Baskets of grain filled the space. It was a storage shed, and the rain trickled through the cracks in the roof and dripped on them as they slept. At 3 a.m. the crows of a rooster, apparently sharing the hut, roused them, and they hustled several complaining chickens out the door. The brothers crept back into their blankets, but from the other side of the bamboo wall they could still hear the chickens that had decided to shelter under the eaves of the house, right next to their heads. Defeated, Ted and Kermit lay under their blankets and watched the rain dripping from the ceiling as the gray morning light slowly took over the sky.

A bamboo house in a Yi village. Photograph by Suydam Cutting, 1929.

The day emerged under heavy cloud cover, punctuated by sheets of pouring rain. The weather was relentless, but the explorers still set off at 6:30 a.m., bound for the mountains. Crossing rivers and negotiating steep ravines, the trail was thick with rhododendron bushes, and the clumps of dark pink, purple, and white flowers doused them with an extra helping of rainwater as they brushed past.

Finally they reached the small town of Kooing Ma. It was two in the afternoon, and they were expecting to see a village bustling with normal activity. Yet not a person could be seen or heard anywhere on the streets.

It felt like a ghost town, with rows of tidy, empty homes without a trace of life in every direction. They began to holler out, to no response. Until they turned a corner to find a rifle thrust in their faces. A man walked forward, clearly the village's leader as he spoke rapidly with the guides at the brothers' flanks. He pointed at Ted and Kermit suspiciously.

"It's your beards," one of the guides explained to Kermit. "They think you are holy fathers."

The villagers thought the Roosevelts were Catholic priests, a profession not respected or welcomed in this region. The Yi had had enough of missionaries and were quick to eject any who came through the area. The guides explained that the beards were only the result of traveling, and that these men were not here to convert others to their faith. They were here to hunt. Finally, after a tense half hour of conversation, the rifle lowered and the group was hesitantly invited to sit.

Finding the right token of appreciation had never felt so important, and Ted and Kermit rummaged through their bags looking for something they hoped would impress the villagers. They had an odd assortment of options, from jewelry to ribbons to weaponry. The hodgepodge of trinkets reflected both the length of their trip and their lack of cultural understanding. Throughout the trip, they'd struggled to guess what kind of gift people would like best.

The first thing they offered the leader was a pair of field binoculars, but he tossed them aside, uninterested. The Roosevelts then pulled out a collection of axes and knives and told the leader to choose what he liked. This met with more favor, and the man's eyes lit up as he browsed the large selection of tools. Kermit noticed, however, that his father, a tall, handsome man, was not so easily swayed by the gifts. He kept a wary eye on the Roosevelt brothers, as if he knew they were hiding their real intentions behind the buffet of shiny weaponry.

After the introductions were made and the gifts hesitantly accepted, the men and women sat down together and began to discuss the panda.

"It was here a month ago," the leader finally offered, "and attacked the apiary" before it was driven away by rifle fire. He explained that the same gun that had been leveled at the Roosevelt brothers upon their arrival had been borrowed from a nearby village for exactly that task.

"Did you ever kill a *beishung*?" one of the guides asked.

"Six years ago," the leader replied, "one was killed in a honey raid."

"Any other time?"

"Never," he responded, and it was clear from his tone that, just like in the neighboring village, the people in this part of China would not intentionally hurt a panda. The rain began pounding against the walls of the house as Ted wondered how to phrase the next question. Hesitantly, he asked if any of the local hunters would be willing to hunt panda with them.

The question was conveyed and Ted and Kermit could feel the eyes of the village leader and his father upon them. The Roosevelts' visages by this point were far from impressive. They did not look like the sons of a famous man, nor as if they possessed any exceptional wealth. Their khaki shorts, once so crisp and neatly pressed, had turned a murky brown and were badly weather-beaten, with loose threads trickling down their calves like jungle vines wrapping around tree limbs. The wrinkles around their eyes had deepened even further, the result of months of squinting in the bright sunlight reflected off snow in the Himalayas, and their beards were wild and thick, streaked now with thin patches of gray. It was an anxious moment; no one could predict how the Yi villagers would respond to the request to hunt a bear that was culturally respected, and for a moment the only sound was the rain outside and the breathing of a dozen men and women within the small room. Finally, the leader replied.

"Yes," he replied, "ten hunters."

The Roosevelts felt a swell of relief. They knew they needed help from hunters familiar with these woods, even if the people they were relying on didn't typically track or kill pandas themselves. It felt like an enormous gift that these hunters would accompany them in tracking an animal they seemed to consider sacred.

What the brothers didn't comprehend was that the villagers had no more experience with the beast at hand than they did. The panda had been spotted infrequently if at all, rarely more than once a year, so they'd agreed to take the rich white men into the woods as nothing more than a lark. No rational being could expect these efforts to result in an encounter with a *beishung*. The local hunters figured it was an easy payday and the risk of encountering, much less killing, a panda was negligible. After only a few days of wandering around in the wet jungle, they thought, these bearded strangers would give up and go home.

The group discussed where they would go, and when they should start in the morning, and then what kind of payment the hunters would receive. They agreed on fifteen to twenty cents a day, depending on experience, with bonuses given if a panda was spotted. The details were arranged by the guides, and then the group shared a meal along with hefty cups of a home-brewed liqueur made from honey, orange peel, and a mix of herbs. Finally warm, dry, and filled with good food and drink, the Roosevelts leapt into their tents, which had been pitched adjacent to the bamboo huts so they could stay as dry as possible.

The rain continued all afternoon and throughout the night while the Roosevelts slept. The ground squished from the mud beneath them, and the canvas was full of leaky holes, but at least there were no roosters to be evicted in the middle of the night. As Kermit sank into a deep, exhausted sleep, a dream he had experienced in many iterations returned. He was in South America once more, fighting a thick jungle, his father weak and fragile beside him. Then a frosty breeze cut through the humid air. The jungle gave way to the rocky, snowy outcroppings of the Himalayas, and Kermit was alone, calling out for his father and then his brother Ted as he stomped through snowdrifts, hopelessly lost. He yelled into the abyss but only silence answered him.

Kermit woke up with a burning fever and full-body aches. "I never had such dreams as I did in China," he recalled. His blankets were damp from a mixture of his own sweat and the steady rain, and his spiking temperature made him shiver violently. He unzipped the tent and looked up at the sky. It was gray and cloudy but at least the rain had stopped. He knew he should rest, but time was running out, the days were short, and the group couldn't waste time waiting for him to recover. He had slept in his clothes, as usual, so there was little to do to get ready for the day but groggily wipe the sleep from his eyes and pat the sweat from his brow. The group, now including ten local hunters, packed up their things early and took off for the mountains.

As they hiked, the Yi hunters looked like mushrooms floating down the trail. They moved light and fast in capes woven from sheep wool and dyed in shades of brown and black. A freezing rain began to fall, and Kermit and Ted were soon jealous of their outerwear. While they were soaked through and shivering in their cotton clothes and patched jackets, the wool capes repelled water, keeping the clothes beneath dry next to the skin.

The trail suddenly narrowed to a slender brown line cutting through the green. The hiking was pure misery. "Drenched by rain and soaked by snow," Kermit wrote in his fever-addled state, "whenever a moment's halt was called, we alternately shivered and panted." As they climbed, patches of snow emerged along the edges of the trail, mixing with the dirt and rain to form a sludge that was becoming increasingly difficult to hike through. They were getting high enough in elevation that the winter snows had not yet melted, and the freezing rain was evidence that the snow was sticking around.

Once again, hunting dogs loudly crashed through the jungle, scaring away all the small game, and probably the larger animals too. Yet the superior nose of a dog could be useful, and it soon became clear the dogs had sniffed out an animal. Ted and Kermit followed the dogs and realized they were chasing a wild pig. "We did not want to shoot pig," explained Ted, "but after a hurried debate we decided it would be better for our future chances, in keeping our hunters keyed up, if we shot game for them."

Chasing after dogs and pigs at a rapid pace was dangerous in thick jungle, especially up in the mountains where the trails ended in precipitous drops and slippery rockslides. The Yi hunters made the route look easy on their light feet, but Ted and Kermit kept falling, grasping at the roots of plants, and twisting their toes in the fissures of a rock face. The sleet began coming down heavily, and since they were as unsuccessful at hunting pig as they were bear, they decided to give up and make camp.

No one listening to the talk around the campfire that night would have guessed the day's activities had been a washout. The Yi hunters and the Tibetan guides got along well, bonding over their shared employment and their distrust of the Chinese, and they drew the Roosevelt brothers into their friendly conversation. Hsuen was especially animated, and even Kermit, brought low by illness, found himself laughing at his stories. "Hsuen's choice of words and *tournure de phrase* were a constant delight and capable of cheering the most gloomy moments."

With Kermit still unwell and struggling, the conversation naturally turned to illness. "The headman's got a ghost," Hsuen explained. "It makes his head itch all the time." Ted reached into his bag and pulled out a small jar of cold cream that he offered to the Yi hunter, explaining that the white cream should be rubbed in his head vigorously to get rid of the fleas.

"There's another ghost," Hsuen translated, "that comes every January."

"What does this one do?" Kermit asked.

"It kills the children."

Ted and Kermit were quiet. "This was undoubtedly pneumonia," Kermit wrote later, "and a more difficult apparition to deal with."

The rain began falling again and the explorers retreated to their tents. Kermit groaned when he saw how much rain was coming through the holes in the canvas. The water dripped steadily, forming puddles at the edges of the seams. They crawled into their bedrolls and tried to get warm, but it was impossible. The wet saturated their skin and the cold sank down to their bones.

The next morning the rain was still coming down. "We were a bedraggled lot even before starting," Kermit described. It seemed that the monsoon season had begun early. The Roosevelt brothers had been told that the rains usually began in April and lasted through September, with typhoon activity peaking during the late summer. They had tried to time their trip around monsoon season, but months of wandering had led them straight into mid-April. The brothers found themselves at the end of a vanishing trail and the last few yards were being washed away by rain.

The explorers slogged through forests and along mountain ridges for days, but not a single mammal could be found. The rain made every minute a misery and they had trouble even staying on the trail. Finally the hunters suggested they travel to a nearby bamboo forest nestled in what they called "the hidden kingdom." The hunters explained to the Roosevelts that the area was full of takins, the wild antelope-like goat the explorers had purchased a skin of many miles earlier and unsuccessfully tried to hunt. Pandas, unfortunately, were only seen "occasionally," but the explorers would take what they could get. They had to try somewhere.

They trudged slowly toward the hidden kingdom. Four Yi hunters accompanied them, along with the Tibetan guides who had traveled more than seven hundred miles alongside the Roosevelts, through war, near-starvation, blizzards, and illness. Kermit noticed that the farther they hiked, even in the rain, the wilder and more beautiful their surroundings became. "At length the valley opened up," he wrote, "and we had easy going. The country became more and more lovely; giant pine trees grew alongside the path."

As they approached a small village, the rain intensified and a "dense fog soon shrouded the valley." Accompanied by the Yi hunters from Kooing Ma, they were welcomed into one of the cabins, cozy with a fire burning in a hearth fitted in the center of the room. Even Ted and Kermit, with their bushy beards, were given a spot by the fire and served strips of roasted wild pig. The conversation turned to hunting, and the brothers listened eagerly as the hunters discussed the prospects.

They all agreed that takin, the mythical goat, would be difficult to find, but panda would be impossible. There were few *beishung* about these woods, and they were rarely seen, not that the hunters had much experience tracking them. Like the villagers in Kooing Ma, the Yi people in this village would not willingly shoot a panda. Soon it became clear, however, that the number of animals potentially nearby was irrelevant.

The monsoon season had started, and the jungle had become impenetrable with rain.

"It's impossible to hunt in this weather," one of the Yi men explained.

"When do you think it will clear up?" Kermit asked through their translator, Hsuen. The hunter shook his head and said that there was no telling, the weather would likely last for months.

"Ask if they can give us just one day," Kermit begged Hsuen.

The hunters and guides debated heatedly while Ted and Kermit looked on. Their future was being decided without them. Kermit poked his head outside and saw a cascading waterfall issuing from the sky. It seemed an entire ocean had been hiding in the clouds that had just been ripped open. The rain was still coming down when Hsuen announced proudly, "They'll do it." Ted and Kermit were grateful, aware that the guide's persistence alone had made the event possible. The explorers went to bed that night listening to the steady patter of rain on the cabin roof and wondering what their last day on the trail would bring.

The next morning, Ted and Kermit awoke expecting more rain—but instead of puddles, a blanket of white covered the outdoors. Snow was coming down in big, soft flakes. The jungle had transformed into a fairy wonderland, and they were surrounded by thick columns of bamboo. The snow settled in the ridges of the bamboo stalks, called nodes, forming an alternating pattern of bright green and snow white that reached far above their heads before fanning out in broad leaves at the top. Kermit described the trees as covered in "conventional Christmas garb," but the beauty of the snow was lost on the explorers, who saw in the fresh coat of ice crystals only memories of their past suffering in the Himalayas. They thought they had finally escaped the ice and sleet, but it had followed them down from the Tibetan Plateau.

It was "a cheerless morning," and a misty rain began to blow across the valley. Four Yi hunters remained with the explorers, and their boots crunched in the snow as they set off toward a corner of the hidden kingdom some seven miles distant. They were headed to a deep ravine that punctuated the valley and was thick with bamboo. For days the thudding of rain had been their constant companion, and now the woods were eerily quiet, the snow dampening the steady hum of the jungle. Even the pack of hunting dogs had been silenced. They had left the group early that morning to chase a colony of rats they'd found hiding in a nearby burrow. The Roosevelts were thankful.

Snow and mist swirled in their hair as they noticed what looked like the dark trunks of a distant stand of trees, stark against the white snow ahead. As they hiked closer, they realized these weren't tree trunks at all, but colossal roots, each one standing more than eight feet tall, and hanging from a maze of curved branches. It was a massive banyan tree with an aerial root system so long and thick that it seemed the wood could shelter half the animals in the forest. They hadn't seen one in hundreds of miles, since they'd first set out from the back door of China's western border and into the curvy hills of the trail.

Ficus macrocarpa, banyan trees are infinite. Growing for centuries, they "walk" through the jungle, stretching their roots far and wide, with some trees stretching four acres long. Their figs provide nutrition, while their bulk provides critical shelter for the inhabitants of the forest. For thousands of years, animals have huddled under banyan trees during times of poor weather. Now it was time for the Roosevelts to cower in the shadows of the ancient giant.

As the group nestled in the safety of the banyan, they imagined what other creatures sheltered here. One animal, of course, was always present in the brothers' imaginations, and as they emerged from the massive tree, sad to leave its protection behind, they suddenly saw it: a paw print in the snow. Ted and Kermit bent over it eagerly, hardly daring to believe it was real. It certainly looked like a bear print with its short digits and long, rounded palm pads. However, unlike a brown or black bear, there was an extra digit, pressed down beneath the typical five toes that Ted and Kermit expected. They could see its unusual indentation in the snow, and their senses prickled with anticipation of what lay ahead.

A light snow was still falling, and Ted and Kermit were anxious to follow the tracks while they could, but the Yi hunters halted. They were speaking in worried voices that Hsuen translated for the Roosevelt brothers.

"They are panda tracks," the guide confirmed, "but they're not sure if we should follow them."

It was a maddening moment for everyone involved. Ted and Kermit, as close as they had ever been to a panda, were ready to rush off into the jungle after the creature. The Yi hunters, on the other hand, were taken off guard by the *beishung* and wanted to turn around. The fool's errand had become all too real. After a heated debate, one of the hunters turned back, unwilling to participate in the hunt, but the rest plodded ahead, steadily tracking the bear through the woods and hoping that the sacred animal was running far away from them. Never had any hunter wanted less to catch their prey. The push and pull of conflicting desires was so strong that the tension filled the air, as palpable as any living being.

The Roosevelts were in the lead, out of necessity, and they could tell from the tracks that the panda was "traveling along in a leisurely fashion." The animal didn't seem to be in any hurry. The bear meandered this way and that, leaving a trail of paw prints in the snow and half-munched bamboo in its wake along the forest floor.

The group followed the trail for two and a half hours, moving quietly though the bamboo and being careful not to rush and scare the animal into running off. The snow had stopped falling but the sky still threatened rain. Then, suddenly, they noticed the slender hoof marks of wild boar mingled in with the panda's wide prints. This was odd. For a mile or more the animals' paths converged and then suddenly separated again. The hunters had no idea what to make of it.

The Roosevelt brothers added their own tracks to the path, although their prints were often just skids in the dirt. This was because the panda followed no path but made its own trail, wandering through the jungle, crossing steep slopes, and crawling over and under slippery, ice-covered logs that Ted and Kermit struggled to balance on top of. "The bamboo jungle," wrote Kermit, "proved a particularly unpleasant form of obstacle course." The stalks of the bamboo were covered in barbed spines, called glochids, that when brushed against left a rash of tiny cuts in the skin. The backs of the explorers' hands were covered in what felt like hundreds of stinging paper cuts. Their hands were so numb with cold, and their brains so buzzed with excitement, that they could barely perceive the stinging pain.

The bamboo was closing in on them. They could see the green columns of the plant on all sides, and above them the feathery tops of the bamboo covered in ice crystals that drooped like massive, sparkling chandeliers. But even underground, where the explorers couldn't see, the bamboo was present. It had formed a crisscrossing network of growths called rhizomes beneath their feet, which would emerge in spring and fill the forest with new tender growth. The hunters were truly surrounded.

Still, the explorers continued. Now that they had found tracks, the Roosevelts were intent on following them, no matter how long it took. The farther they walked, however, the warmer their bodies became. As they loosened their jackets, they realized that it wasn't just exertion making them sweat but that the sun had emerged from the clouds. Ted looked up at the sky angrily. They hadn't seen the sun for a solid week and it seemed a cruel twist of fate that it would decide to make an appearance now, when they had finally found a panda. They hurried, desperate to beat the rush of warm sunlight that would melt the precious tracks leading them onward. They needn't have worried. Perhaps witnessing their pleas, the sun tucked back in behind the clouds.

Even if the snow had melted, the signs of the panda were becoming more prominent. On one side of the trail a "nest of bamboo" perched in the bushes, while on the other Kermit spied claw marks in the bark of a tree. Then there was the scat, obviously the panda's given the amount of bamboo in the droppings. It seemed they were finally going to catch up with the bear when they heard a yowl from one of the hunting dogs below them in the valley. The timing could not have been worse, and Ted cursed the dogs angrily and wished they were anywhere else. Still, they drove onward, feeling a rush of anticipation as they hiked.

The eager feeling of expectation was not unlike roaming their childhood home in Oyster Bay in pursuit of their father. Climbing the stairs, they would hear their father rattling around in one of the bedrooms. The bear was close. Each creaking step on the old wood floors brought them nearer, their hearts pumping furiously with adrenaline as they inched toward the man they always wanted to please, and the person they looked up to most in the world. Every step since the day their father had died, the brothers had been searching for the "old lion." They looked for him in relationships and in jobs, but most of all in places like the Himalayas and the Tibetan Plateau, where the wilderness transformed them back into the boys they had once been. "I seek my father in the wild places," Kermit once wrote. Nothing could be more remote than the jungle they trekked through now, and nowhere had their desire for the hunt been so strong.

A strange "clicking chirp" sounded from the trees. Kermit looked in the direction of the noise, wondering what it could be. "It might have been a bamboo snapping or the creaking of the interlocking branches of two trees swayed by the wind," he postulated. His instinct, however, told him something different. There was an animal there; Kermit was sure of it. That gut feeling that his prey was close filled his belly and flooded his senses.

One of the Yi hunters moved silently through the bamboo and then, when he'd gotten forty yards ahead of Kermit, signaled enthusiastically for Kermit to follow. As quietly as he could, Kermit joined him. Before them was a giant dragon spruce, its bright-blue needles contrasting with the rich green of the bamboo on either side. Looking up, Kermit was startled to see a large hole in its trunk. Suddenly, the head of a panda bear emerged.

The bear was exactly like the illustrated plate of the animal the Roosevelts had obtained in France, with a head covered in white fur and "black spectacles" around the eyes. It moved sleepily from the tree, as if not completely awake, and sauntered out toward the bamboo. One of the guides was gathering Ted, who had explored in the other direction for a few minutes, and Kermit waited for him impatiently. "Though in reality it was only a short time," he recalled, "it was a nerve-wracking wait."

LAND OF THE YI

As Kermit watched the panda, he felt a strange sense of detached reality. "He seemed very large, and like the animal of a dream," he wrote. After a thousand miles and five months of searching, it seemed impossible that this bear in front of him was real and moving through the jungle just thirty feet away. Kermit followed the panda as it eased through the bamboo, cautious of every footfall. The *beishung* seemed like a mirage floating in front of him, one that he worried would "vanish like smoke in the jungle."

Ted finally caught up with his younger brother, and the two men raised their rifles. This was the moment of truth. "We fired simultaneously at the outline of the disappearing panda," Kermit wrote, although few of his friends and acquaintances would believe this part of the story. It seemed too convenient.

The poor animal, still dazed from sleep and now wounded, turned toward the Roosevelts, and the brothers looked the object of their desire directly in the eyes. For a moment they saw themselves reflected in its dimming gaze. Then the bear floundered in the snow, surprising the brothers by showing no sign of aggression or challenge. "Although the Lolos had all along told us that *beishung* was not an animal to be feared," explained Kermit of the chase, "we used caution in following the trail." No special care was needed, and the hunters were in no danger whatsoever. It was only here, at the end, that the brothers realized they had been wrong and that the bear wasn't the wild, bellicose predator they had expected.

Ted and Kermit once again raised their rifles and shot, this time at a distance of only five or six feet. Stumbling and fighting for its life, the panda fell over but then regained its footing and took off deep into the safety of the thick bamboo jungle. The bamboo had nurtured the bear, provided all its meals, allowed it to communicate with others of its kind, and formed its home. The plant had even fought against the Roosevelts, slicing the hunters' hands as they hiked through its evergreen walls. But there was nothing that could be done now. Time was running out.

"We knew he was ours," wrote Kermit. But he wasn't. He belonged

THE BEAST IN THE CLOUDS

to China, to the bamboo forests, and to the people who revered him. He was not a prize to be won or a dream to be fulfilled. The most peaceful creature in the forest was in mortal peril, and his pursuers followed him mercilessly.

The brothers traced the trail of blood, each step bringing them closer to their fate. The drops of red were easy to spot contrasted with the bright green of the forest, and they moved eagerly, the momentum pushing them forward. Chasing an animal in the woods was thrilling, and they became boys again, moving swiftly through the sea of green. Ahead they saw only success, achievement, and fame. The dark reality of what their accomplishment would mean was a phantom darting in front of them, obscuring their vision, and as dangerous as a beast in the clouds.

The panda stopped and fell to the forest floor, and red blood began to stain the black-and-white coat that hundreds of white men had pursued across the vast continent, dragging along with them the inhabitants of so many hidden kingdoms. Like Ted and Kermit's father a decade earlier, the giant had fallen. The bear hunt was over.

Yi hunter and Ted with the panda directly after the hunt. Photograph by Kermit Roosevelt. Courtesy, Field Museum. CSZ67964.

CHAPTER 12

THE HALL OF ASIAN MAMMALS

For months the brothers had pursued the panda across a thousand miles of ice, snow, jungle, and the wind-beaten rock of the Tibetan Plateau. Yet when they finally accomplished their goal, tracking and killing the bear, they turned despondent. It was the sensation of getting everything you wanted then suddenly being unsure if it was what you truly desired in the first place.

"We had hunted hard and long," Kermit wrote, "usually in the face of every adverse circumstance." Yet no words of joy and no boast of accomplishment entered his accounts of the days ahead in his journal. Instead, the Roosevelts noted that the panda "was not what we were expecting." They knew their feat would be plastered on the front of newspapers, discussed around drinks at the explorers' club, and celebrated wherever they went, but their feelings were dampened by the reality of what they had discovered in the woods. The panda was not a killer who rampaged villages in Japan, or a mighty foe feared by those traveling in the Arctic. Instead, they had found an animal who challenged their core beliefs about big game hunting.

After the panda had been shot, and with a massive effort, the Yi hunters had hauled the animal out of the jungle. At first, the men and women protested bringing the panda into their village. They could not imagine the carcass of the sacred animal entering the place where they lived. The Yi villagers relented only when they realized the Roosevelts would have to prepare the animal in the mud and snow on the outskirts of the town. "A deeply interested group surrounded us at our work," Kermit remarked, as they began the arduous task of measuring, photographing, and describing the animal before laying out the hide, placing it fur down, and then rubbing salt into the exposed skin and paws. All this had to be done before they could crate the panda for the journey home.

They made detailed notes, filling dozens of pages, concerning habitat and diet. From their minute observations, they remarked that there was "no trace of the panda having varied its bamboo diet." It was clear from their travels, and from examination of the panda's behavior, that the animal resided only in dense bamboo jungles like this one, and they knew from tracking it that it liked to browse through the bamboo slowly, eating constantly as it traveled.

They skinned the panda gently, sliding a knife straight up the animal's belly until they reached the base of the neck. Every motion was smooth and cautious, and they treated the dead bear as respectfully and tenderly as possible. There was a finality to each of their movements that neither brother had experienced previously during an expedition. When Kermit had hunted in Africa with his father, for example, he had thought it possible he might return one day, tracking lion and elephant once again in his lifetime. Yet from the moment the panda had fallen in the woods, Ted and Kermit knew that this was it, the only time in their lives they would hunt this black-and-white bear. The journey had changed them, and they were no longer the hunters they had once been proud to call themselves. The bear had stirred feelings in them that were new and unfamiliar.

Kermit took to his field journal, attempting to explain that which he'd always found so inexplicable in describing why the panda was different. "He's a gentleman," was where he landed, the words still falling short.

Coming at the end of a long trail, full of the unexpected, nothing had been more startling than the panda itself, a peaceful creature who lived alone, far from humans and even others of its kind. Unlike brown and black bears, frequently portrayed as overtly aggressive in their interactions with humans, the panda had been gentle, solitary, mild. The hunt had left the Roosevelt brothers paralyzed. They knew immediately, from the moment the panda fell, that they never wanted to hurt another of its kind. "It's not a savage animal," Ted would later explain. The panda had permanently altered their sense of purpose, in ways that were about to reveal themselves.

During their previous hunts in China, they had developed a ritual. Sharing the meat of the animal they had killed with the hunting party provided a sense of closure and permanent connection, but with the panda, none of the men and women "would touch a morsel of the flesh." Ted and Kermit felt the same way. Instead, the head of the village ordered a sheep to be slaughtered, insisting to the Roosevelts that he would pay for the substituted meal himself.

Meanwhile, the villagers held an "all-embracing ceremony of purification," whose purpose was to "cleanse the house and its surroundings from any shadow that the death of the giant panda might cast upon it." Religious figures had been sent for and they began work immediately, completing the ceremony only after the Roosevelts were gone. They were purging the village not only of the panda's sad fate, but of the presence of the Americans as well.

The feast began and wine was passed around, along with bowls of rice and boiled meat. Ted and Kermit stayed up late, speaking of the panda, and sitting in tanned leather chairs around a roaring fire that was fashioned in the center of a bamboo hut. The Roosevelts were particularly taken with the matriarch of the village, an older woman named Vooka. Superficially she resembled many of the women they had encountered on the trail, yet there was something different about her. She seemed taller than she was, her voice charming, and despite her health problems and severely injured eye, she had a quiet dignity about her that reminded the Roosevelts of grand society ladies.

Upon meeting Vooka, the Roosevelts had presented her with "a gold ring set with turquoise" but the old lady had refused, waving it away. She thanked them for their generosity but protested that she was an old woman, with "no use for gewgaws." What she wanted was not a gift, but conversation. Over bowls of mutton and cups of corn wine, and with the aid of Hsuen, their translator and guide, they recounted the trip once more. The village, the panda, and then the rest of the world. Vooka was curious about everything, asking the Roosevelts about their lives in the United States and what their travels were like.

The moon was shining overhead, only partially obscured by clouds, when the bowls were empty and the wine finished. The Roosevelts climbed into a neighbor's hayloft, completely exhausted but grateful. They piled the blankets on top of themselves and let their minds fully relax for the first time in months. The journey was over, but even as they slept, "black and white *beishungs* were never long absent" from their dreams.

The next morning, Ted and Kermit paid the wages and bonuses that were owed, and then showered the village with presents. While Vooka had snubbed the jewelry they'd first offered, the Roosevelts now knew the way to her heart, and presented the matriarch with a .32 Colt automatic pistol. She was overwhelmed with gratitude and told the Roosevelts that she was "really embarrassed" and wished she had a gift for them, but her people had "lost most of their family treasures when they were driven into this valley from their old home." The Roosevelts, however, had already received their heart's desire, the large crate that followed them home as proof of their skill as hunters, scientists, and even, strange as it seems, conservationists.

"When we bade her goodbye," wrote Kermit of Vooka, "she had her family and many of the guests of the previous evening around her. I believe they were sorry to have us leave, and I know we were sorry to go." The Roosevelts described Vooka as a "kindly old woman." "It seemed almost unbelievable," wrote Kermit as they left the land of the Yi, "that these were the savages we had been told would rob us."

THE HALL OF ASIAN MAMMALS

The tender feelings stirred by the final good-bye would not last. From the moment the brothers left panda country with their black-and-white prize, paid for in blood, a shadow fell over their fates.

Immediately following the panda hunt they were struck by illness. They were traveling to the city of Yunnanfu, more than fifty miles distant, to meet Jack and Suydam. After three days on the trail, Kermit felt his stomach clench and soon his whole body was in spasms and pain. Ted wasn't doing any better. A cut on his leg, received while chasing the panda through bamboo, became infected and the bacteria spread in angry red lines that crisscrossed up to his torso. His head ached with fever. Still they trudged onward to meet up with the rest of their group. In their misery, they donned large straw hats to shade themselves from the warmth of the early May sunshine. "It seemed impossible," wrote Ted, "that only three days before we had been in rough forests waist-deep in snow."

In their weakened state, progress was slow, and the Roosevelts still had no means to communicate. Although they had left the borders of Lolo land, they were unable to get news. As in most of the wilderness they had traversed, no town they passed through had a telegraph and it took months to get mail delivered. They hadn't received any letters since December of last year, when they'd started their journey, so the world they were returning to was unknown.

The brothers crested a ridge and then saw the sparkling waters of Lake Yunnanfu below. It had taken a week, but they had managed to get to the city where they would gather their party back together. Suydam and Jack would be there, waiting for them, and they could send a message to Herbert, care of the Cunninghams in Tatsienlu, to update him on their successful hunt and let him know where to find them. They couldn't know that Herbert had only just left the Kingdom of Muli a few weeks earlier and was still wandering the high peaks of the Himalayas, a full month behind schedule.

Delays didn't matter to the Roosevelts any longer. They could rest in Yunnanfu, recover, arrange to ship the panda back home, and then forge onward, this time with a newcomer to the group. A year earlier, they had arranged to bring another scientist, Russell Hendee, onto the expedition as they extended their journey into Southeast Asia. Hendee had traveled to Indochina, his first time visiting Asia, at the same time the Roosevelts were starting their expedition, in December 1928. Hendee was collecting birds, reptiles, and small mammals with a team of scientists led by Harold Coolidge Jr., a zoologist at Harvard University, also funded by the Field Museum as part of the Roosevelt expedition.

From left to right, Theodore Roosevelt Jr., Suydam Cutting, Culver B. Chamberlain (U.S. Vice Consul of Yunnanfu), Kermit Roosevelt, 1929. Photograph by Wide World Photo.

THE HALL OF ASIAN MAMMALS

Hendee was a relatively young scientist, just thirty years old, and still trying to prove himself in the field. He had trained at the University of Iowa before taking part in an expedition to Alaska for the Colorado Museum of Natural History in 1921, and then to Peru for the American Museum of Natural History in the mid-1920s. In Peru he'd fallen in love, marrying a Peruvian woman, and they settled down in Brooklyn. The newlywed could have stayed with Coolidge and his familiar group of scientists—it might even have been the prudent thing to do—but Ted and Kermit had asked him to join them in May, once monsoon season started, and he knew how much the Roosevelt name could help his career.

Now that the Roosevelts were done with panda hunting, the next part of the trip would mimic Herbert's slow and easy pace, as they identified unusual small animal species in a part of the world that few scientists had studied. After everything they had been through in the past six months, a leisurely trip through warm jungles, listening to the peaceful melodies of birdsong, sounded like heaven to them.

The brothers had barely arrived in Yunnanfu when news began flowing in. After months of being cut off from the world, the sudden influx of letters and cables was overwhelming. Ted sorted through the stack of papers. There were family celebrations he had missed, political gossip, and news of a new president, Herbert Hoover, who had taken office. He glanced over at Kermit and saw that his brother was visibly upset. Unlike Ted's mild stack of news, Kermit's was more troubling.

"Look," Kermit said, handing Ted a pile of papers. It was his shipping business, Ted realized as he began to read through the cables, each marked *Urgent*. Kermit had perhaps never been a dedicated businessman, thus his willingness to leave his company for a year, but now he was in desperate trouble. The business was heading to bankruptcy. His only hope was to jump on the next boat to New York and try to save it.

Ted was uncomfortable. They depended on each other in ways that only close siblings can truly understand. When he was without Kermit, Ted didn't feel fully himself, particularly in the woods. After months of constant togetherness, companions in every hardship, Kermit's forced exit felt like a physical tear in his flesh.

The moment Kermit left to make his way home, Ted felt himself unraveling. He wasn't prepared to be alone. A few months later, in a letter to a friend he would describe the trip as the "most perilous time of my life." Yet the danger he mentioned focused mostly on Kermit's absence. "On this last expedition he had to leave two months before the end, and I missed him greatly."

Onward Ted went, even though he knew Kermit was truly the stronger outdoorsman of the two. His body ached from months of sleeping on the ground, repeated illness, and hard climbing. His figure was lean and muscular, but also thin, as the months of trekking had wiped him of his fat stores. The sun had turned his skin a leathery brown, and his beard, which he still hadn't shaved, was flecked with gray, each silver strand epitomizing an adversity found on the trail.

Possessions and trinkets seemed less valuable to Ted than they once had. He gave his precious rifle away to Hsuen, his favorite guide, and lamented that the group of explorers was breaking up permanently. He became deeply sentimental. "Together we had shivered in the bitter winter cold of the high mountains," wrote Ted, "and sweltered in the damp heat of the semi-tropics. Together we had passed through troubles ranging from lost mules to bandits. Now in all probability we would never meet again."

After Kermit left, Ted and Suydam took a train to Saigon, where they planned to meet Russell Hendee and begin collecting animal samples for the museum. A train felt like luxury to the explorers, who were happy to be separated from their mules and to give their sore, callused feet a rest. Hendee, however, had to take a very different route from his team's current position in Laos.

Since it was the rainy season, the young scientist would be embarking on a "thousand miles of downstream travel mostly by small native canoes and rafts." He expected it would take about a month to reach Saigon. With all the confidence of youth, he wasn't worried. In fact, the only part of the trip that concerned him was the rapids that twisted the Mekong River in sharp curves and over death-defying rocky cliffs leading to Vientiane, the capital of Laos.

For a small boat the river was all-powerful, the rushing water and rock merging into a deadly combination that sent even experienced rafters flailing down heart-poundingly steep seventy-foot drops. Aware that his life was at stake, Hendee had hired a river guide, a young man who was also conveying a French missionary downstream. It didn't matter who was paying the bills; there was no room for passive riders in the canoe—everyone held a paddle and was expected to help navigate the fast-flowing water as instructed. "Left!" came the yell from the river guide, as a swell of water rose to meet them. Immediately the men thrust their paddles into the river and paddled madly against the river's fearsome grip. The boat spun through the current wildly until they had made it past the danger.

On the other side of the cliffs, the water was surprisingly calm, and the river lapped gently against the sides of the boat as if it were a gentle pet curling up at the toes of its master instead of the fierce beast that had just attempted to kill them. Hendee wasn't fooled. Soaking wet and gasping, the zoologist felt a surge of relief. The water was unusually high, as it had been a wet summer, and he helped drag the canoe to shore where they would camp that night. They had made it safely down another stretch of river and now they were just two days from the capital.

It was early evening and the bugs, always present, began swarming. The mosquitoes bred in the small, stagnant pools that peppered the water's edge, and the presence of humans, like any warm-blooded mammal, was a beacon that drew them in. Hendee swatted at the insects, drew closer to the fire, and then finally gave up, retreating under the mosquito net he'd packed for the trip. Insect bites were just part of the job, like sketching birds or estimating the height of banana trees. Hendee had been plagued by mosquitoes since he'd started down the Mekong River, just like every other traveler it seemed, and had become used to the itchy bumps that dotted his arms, legs, and neck.

In one evening, everything changed. Hendee went from perfectly healthy to deathly ill. It started when his body began to ache uncontrollably. Then the fever began, the heat rolling off his skin in waves of agitation and shivering so intense he could barely stand the sensation. To the river guide, the illness was clear. Hendee had malaria. The guide wasn't unduly alarmed. Lots of people got malaria. They rested and drank water, and most of the time they recovered. Except when they didn't.

By the time the guide managed to bring Hendee to Vientiane, the young scientist was fading in and out of consciousness. *Plasmodium falciparum*, the parasite that causes cerebral malaria, was causing his brain to swell. His fevers were so high that he had become delirious. He'd gotten the disease through an infected mosquito, and although the hospital had a supply of the only known treatment, a drug called quinine that was extracted from the bark of a type of cinchona tree in South America, the parasite had developed resistance to the widely used medication.

In his hospital room, Hendee could barely communicate. The only thing he knew was that he desperately wanted the pain to stop. He looked out the upper-story window at the beautiful city of Vientiane but saw neither the celebrated golden temples nor the massive Buddha statues the city was known for. He was in agony and through his open window he saw only one thing: a way out. He stood on the ledge a minute and felt the rough wood on his bare soles. It was as ragged as a torn page from a field journal, sending a row of splinters into one heel. With a leap, the young man plunged to his death.

On the sidewalk below, hospital workers saw Hendee immediately and moved toward his body, but it was too late. There was nothing they could do for the young man. His pain and misery had ended the minute he hit the paved street.

Arriving in Saigon, Ted and Suydam were shocked by the news of Hendee's death. The scientist had been young and healthy, and they couldn't believe he was gone so quickly. It had only been a week since he first fell ill. They made the sad preparations of sending cables to Hendee's mother, his new bride, and his colleagues back home while the hospital prepared for his burial. In a small graveyard in Laos, a stone marked Hendee's final resting spot. It read:

In Memory of Russell W. Hendee
Expedition of the Field Museum of Natural History Chicago
Died Vientiane June 6, 1929

Ted and Suydam would never see the grave in person. One evening, the eldest Roosevelt brother began perspiring so heavily that his sheets became slick and transparent. It was a fever, and as he looked across at Suydam in his bed on the other side of the bedroom, it was clear that his friend was suffering as well.

Here was the same disease that had so recently killed a member of their party, and fear racked their bodies along with the fever and chills. Just as for Hendee, the parasite had become resistant to quinine. Chloroquine, a drug that would prove effective against malaria, was still five years away from discovery. Their only hope of survival lay in their own battered immune systems.

Two days later, Ted and Suydam were admitted to the hospital in Saigon. Like Hendee, they were barely conscious most of the day; the parasite had laid them low with high fevers, filling their brains with bizarre hallucinations and making them delirious. In addition, the doctors found that both men had dysentery, a gastrointestinal disease caused by either bacteria or parasites. Here was dangerous news. Instead of drawing upon a healthy immune response to ward off the *Plasmodium* parasite, the two men clenched their bellies as they experienced massive cramping, nausea, vomiting, and diarrhea.

Ted could take only a few delicate steps in his hospital room before collapsing. He was so weak and miserable that he couldn't have flung himself from the window even if he'd wanted to. He was losing weight rapidly, his kidneys were shutting down, and just days later his brain was functioning so poorly that he could no longer speak or feed himself.

His fever climbed to 105 degrees Fahrenheit, a dire medical emergency. The end seemed near, and Suydam, miserably sick himself, could offer nothing to help his friend.

Malaria was endemic to much of Southeast Asia, along with other regions of the world, but in Vietnam the disease had proved especially deadly after the country began to increase its rice production. These farms, typically located on the edges of large cities, were a perfect environment for mosquitoes, containing what the insects liked best: standing water. As mosquito populations grew, they transmitted malaria across countryside and city alike, resulting in thousands of deaths.

The hot, humid June temperatures in Saigon proved a happy environment for the mosquitoes, but a miserable one for Ted and Suydam. As the two men struggled to survive, Eleanor, Ted's wife, was making her way to his side. She had always planned to meet him in Saigon, but hadn't expected her husband to be rendered unrecognizable at the end of the trail. Although she'd been prepared by Henry S. Waterman, the American consul general in Saigon, who met her boat on the dock, she was still shocked by what she found.

"Never in all my life have I seen anyone who looked so sick," she wrote. "Not even Ted when he was gassed in 1918. He had lost forty-two pounds." Her husband was a frail 102-pound skeleton who could no longer remember how to open doors and refused to eat. He was released by the hospital and allowed to stay in a hotel, but his fevers were still dangerously high and his delirium was obvious to Eleanor. It was terrifying.

While Suydam recovered enough to travel back to the United States, Ted was stuck in Saigon a little longer. Someone needed to coordinate the end of their expedition and ensure that the samples they'd collected in Indochina and Vietnam would make their way back to the Field Museum. With Ted weak and wretched, the work was left to Eleanor, who had to sort and ship the samples, organize the equipment, and pay the bills. By late June, the last Roosevelt left Asia for home.

The life the Roosevelt brothers were returning to had altered. It was summer 1929, months before the stock market crash that would usher in the Great Depression, but both men were on the brink already, along with much of the country. Newspapers across the country were heralding the Roosevelts' successful expedition, with the *New York Times* describing the animal as a "rare beast of the Tibetan border" with "strange markings," and calling it "the first giant panda ever shot by a white man." There was no outcry of barbarism or cruelty, as one would expect today; instead the papers were celebratory. Here, finally, was proof of the bear whose existence had long been doubted. The panda was real, with a skeleton that scientists could finally study, and the Roosevelt brothers began to receive letters requesting the details from naturalists across the globe.

"Wild animals," President Theodore Roosevelt had once explained, "only continue to exist at all when preserved by sportsmen. The excellent people who protest against all hunting, and consider sportsmen as enemies of wild life, are ignorant of the fact that in reality the genuine sportsman is by all odds the most important factor in keeping the larger and more valuable wild creatures from total extermination." Such was the philosophy not just for hunters of the day, but also for scientists.

Excitement over the Roosevelt expedition only grew as the months passed. In December 1929, the roads in Chicago were covered with a slick coat of ice but the weather wasn't keeping people at home. More than three thousand people came to hear Kermit speak at Chicago's Field Museum, their figures snaking into a long line that wrapped around the block. Only half that number would get in; the rest would be turned away once the room reached its capacity.

Kermit's mind, however, was seven thousand miles away from Chicago and seven months past, back in China, after the panda had fallen in the woods. The passage of time had crystallized his regrets. "I'm not myself," he'd confessed to his wife. It was an understatement.

The unintended consequences of the expedition were beginning to show themselves. Big game hunters were now calling the panda "the most challenging animal trophy on Earth." It was clear that the maps the expedition had made, marking where the panda could be found, including its habitat and diet, were going to be used for a deadly purpose.

Just three months after the Roosevelt brothers returned, five new hunting expeditions were making their way to China.

Not everyone pursuing the panda intended to kill. There were some explorers who, now that the panda's existence and location had been marked for Westerners, dreamed of bringing one back alive. Such an exhibit would be the crown jewel of any zoo, able to draw in throngs of visitors. There was no limit to what a zoo might pay for such a prize. As difficult as the endeavor would be, the Roosevelts had made the route clear, and while an adult would be impossible to capture and ship back to the United States or Europe, a small cub that could be hidden from the Chinese authorities, and toted in a bag like a puppy, would be far easier to smuggle out.

While "panda-mania" began sweeping the United States, the Roosevelts grappled with their own role in unveiling the animal. As Kermit prepared to take the stage in Chicago, he knew that what the audience was after were stories of adventure, not tragedy. The Roosevelts had hastily published a book in late 1929 called *Trailing the Giant Panda*. Each sentence was ripped from their field notebooks, but erased were the moments of mortal reflection, guilt, and even acute fear. It was a tale of adventure, not introspection, and Kermit knew he had to parrot its tone as he stepped up to the podium.

Elsewhere in the museum, the panda was being thoroughly studied by biologists and taxonomists before an exhibit was prepared. In the Hall of Asian Mammals, a mural of an unnamed mountain would be framed by thick bamboo hung in the background, while gray boulders and fake bamboo occupied the foreground. In the middle were two stuffed panda bears munching on bamboo. One was the skin the Roosevelts had bought in Complete Heaven, the other was the animal they'd killed in the land of the Yi. When the exhibit was unveiled in 1931, it seemed that everyone wanted a piece of this legendary but no longer mythical bear.

Fashion designer and socialite Ruth Harkness was among them. Her husband, Bill Harkness, had dreamed of bringing a panda back to the United States himself before succumbing to throat cancer in Shanghai

THE HALL OF ASIAN MAMMALS

The panda exhibit in the Field Museum. Courtesy the Field Museum, photograph by John Weinstein, 2007.Z94466_13d.

in 1936. Ruth decided to take up her husband's cause and that same year traveled to China with Jack Young's younger brother, whose English name Quentin was a tribute to Ted and Kermit's younger brother who'd died in World War I.

Harkness managed to smuggle a six-week-old cub, weighing just four pounds, twelve ounces, back to the United States after paying China an "export tax" of $45. She named the female bear Su-Lin, after Jack Young's new wife, and checked into the luxurious Biltmore Hotel on Madison Avenue with the small creature tucked into her coat. There she bottle-fed Su-Lin with a mixture of powdered milk, syrup, and cod liver oil, and began auctioning off the panda to the highest bidder. She expected at least $15,000 for the animal, thinking the New York Zoological Society would be the lucky beneficiary, but the New York zookeepers were skeptical. They worried about the cub's health generally and the poor diet it had been subjected to since its arrival in the States. The Brookfield Zoo in Chicago was not as picky, and Harkness sold the panda cub to the institution for $9,000.

Before bringing the cub to Chicago, she asked Ted, Kermit, and Ted's teenage son Quentin to see the cub first. It was a bizarre experience for the veterans of the panda trail. The last time Ted had seen a panda was now seven full years earlier, deep in the bamboo jungles of the Yi people, a far cry from a socialite's hotel room where he held the black-and-white bear in his lap. In a letter to a friend, Ted explained how he wanted to protect the tiny creature. "If it was up to me," he wrote, "it would never enter a zoo."

The expedition was behind him, he'd survived the trail, and yet the years of distress that followed had transformed his perspective. In 1933, Franklin Delano Roosevelt had been elected president of the United States. Ted knew that this was the nail in the coffin of his political aspirations. He'd floundered uncomfortably in foreign posts for years, hoping to make a comeback in American politics. Now it was all over. FDR's rise to office made it untenable, if not impossible, for Ted to work for the government again.

Ted had been appointed governor-general in the Philippines after the panda expedition. With FDR's election, he lost his position. As he was preparing to leave his post, a reporter asked him to clarify his familial relation to the new president. "He's my fifth cousin," Ted tartly responded, "about to be removed."

To make matters worse, the Roosevelt brothers were barely speaking to one another. Kermit had become an alcoholic, his shipping company was still bleeding money, he was cheating on his wife, and he'd gone so far as to meet with FDR in person. To add to the treachery, Kermit had written the newly elected Democrat a gushing note, explaining that although he'd supported Republicans all his life, "I am tremendously relieved and pleased that you were elected." Kermit didn't care about politics—the note was mere flattery before he asked FDR for financial help—but it was still more than Ted could bear from his lifelong compatriot. Since coming home from the panda hunt, Ted's and Kermit's lives had shattered into pieces around them, and it seemed impossible there had once been a time when they'd spent every minute of every day together, exploring the wilderness of Tibet and China.

THE HALL OF ASIAN MAMMALS

Though they were reunited now in Harkness's posh hotel room, the mood was cheerless. The cub cradled in Ted's arms was proof of their carelessness, and the unintended consequences of their ambition. One of the other visitors lightheartedly asked Ted if he thought the cub would one day join the Roosevelts' trophies in the Field Museum.

Ted replied irritably, "I'd just as soon think of mounting my own son as this baby."

Yet the Roosevelts had wrought this future, one where Westerners would build such winning commercial enterprises around the panda that their deaths and the capture of their young would be fueled by excessive sums from the extremely wealthy. Ted and Kermit knew they had unwittingly forced a turning point. Hunting and conservation had always been inextricably linked. Now the knot was loosening, as it became clear that human activity could impact the biodiversity of species. The idea that people could wield this kind of power had once been unthinkable, but now hunting trips could no longer be justified as saving the lives of endangered animals. "We were responsible," Kermit wrote in his diary, the depth of his actions finally clear.

The young bear struggled at the Brookfield Zoo. No giant panda had been kept in captivity previously, and the zookeepers quite literally had no idea what they were doing with the new arrival. Just a year later, the sweet cub died from pneumonia. So little was basic panda biology understood at the time that it wasn't until after her death that zookeepers learned that Su-Lin was male, not a female as they had all presumed. This is because external genitalia in the giant panda don't appear until the cubs are several months old, with male and females appearing similar at birth. While the zookeepers in Chicago could have consulted naturalists at the Field Museum who would have clarified Su-Lin's gender, they instead wallowed in their limited understanding.

For a flat fee of $150, the Brookfield Zoo shipped Su-Lin, along with a cadre of unrelated dead animals including a kangaroo, a rat, and a monkey, to the Field Museum.

Just as the Roosevelts had done in a dense and rainy bamboo jungle

Su-Lin with a young visitor to the Field Museum, 1938. Courtesy of Field Museum. Z80904.

years earlier, the animal's skin was slit, its blood drained, its skin dried, and its insides stuffed before it was placed on display. Her death did not deter others: other panda cubs were now regularly being snatched from China and sent to zoos in the United States. Another young cub, this one given the name Pandora, had just arrived in New York.

In Chicago, the exhibit was given a placard with the bear's name, Su-Lin, and described as a "Popular Panda." She was placed alongside the Roosevelts' panda exhibit in the Field Museum, the dead animals speaking loudly to those who would listen although they could no longer utter a sound.

CHAPTER 13

THE SUMMER WHITE HOUSE

In Tibetan Buddhism, redemption already exists in every single person; it only needs to be awakened.

Suydam found his path back to Tibet printed on an old slip of paper. It was, thin, worn from age, and crinkled from being stuffed in wallets, but its ink was still vivid and readable. A year was written on its corner: 1929, the year of the trail. It was a period in Suydam's life that had made an indelible impression, altering his world in profound ways, but this small slip of paper was now, two years after it had been issued, poised to open a door into his future.

Tibet was not a country that welcomed foreigners easily. Anyone wishing to visit had to apply for a permit, which was rarely granted. Lhasa was called "the forbidden city" by Westerners because its borders had been so rarely crossed, and for good reason. In 1904 a group of British soldiers, led by Sir Francis Younghusband, entered Tibet armed with rifles and Maxim machine guns and quickly defeated a force of roughly six hundred Tibetan soldiers who wielded only swords and flintlocks. The British forced Tibet to sign a treaty allowing England to trade freely in their country and, adding insult to injury, requiring the Tibetans to pay a fee for the privilege. Two years later the treaty was mostly repudiated, but Tibet was understandably not receptive to European visitors.

By the 1920s, Westerner applications to visit Tibet had become far more common, with nearly all rejected. Few scientific expeditions, no matter how worthy, were seen as worth the risk to the Tibetan government. When the Roosevelts came along with their charm, famous name, and bold proposition to find the elusive giant panda, their expedition became one of the few to be issued a permit they would in fact never use. Instead, the Roosevelts' adventure stuck to the border regions between Tibet and China, where they were often unsure which country's terrain their footsteps were treading. Now Suydam had the precious piece of paper in his hands, and it was still valid.

While Ted and Kermit had no interest in going back to Asia and weren't tempted by generous offers of hunting and scientific expeditions that frequently came their way following the panda hunt, Suydam was different. He longed to see the Himalayas again, and especially the Tibetan people. It seemed that every hunting expedition heading to China wanted to include him in their party, along with Jack Young. There was more work being offered than the two men could possibly take, and because they had been part of such a prominent and successful expedition, large sums of money came along with those invitations.

But Suydam had no wish to return under the conditions those third parties' offers called for. Hunting in general had lost its appeal for him, much less a command to stalk and collect more pandas. But when Ted and Kermit told him of the permit to enter Tibet, one with no strings attached or demanding external funders to please, Suydam's imagination was allowed to wander again. He chose to study plant life only.

"I sought no hide-and-seek, no peephole adventure in Tibet," Suydam explained. "My aim was to *study* the country, to see the nomads, peasants, ruling classes, to learn something about that curious offshoot of Buddhism known as Lamaism, to have a glimpse of the arts and crafts, to collect specimens of the flora. And most of all, I hoped to see the country's mysterious ruler, the Dalai Lama, face to face."

Suydam had lofty goals for his visit, more than could reasonably be accomplished in one trip. His first visit to Tibet led to another in 1935, and then a third in 1937, this time with his wife. Newspapers would call him the "first white Christian ever to enter the forbidden city of Lhasa." His wife, Helen McMahon, would become the first white woman to meet the Dalai Lama.

The women in Tibet reminded Suydam of those the expedition had relied on as guides on the trail of the panda. "Freedom was part of their tradition," he wrote, "little girls in Tibet being as welcome as boys. Every Tibetan woman has the right to select her own husband, run her own household, and to own property."

Suydam, as assertive as ever, managed to strike up a correspondence with the Dalai Lama on his first trip to Tibet in 1931. They shared their thoughts on economic conditions in the country, discussed how to import wool into the United States, and bonded over the lovable nature of dogs in letters sent back and forth for years, until the ruler's death in 1933. At one point the Dalai Lama even sent Suydam a pair of Lhasa apsos, a small Tibetan breed of dog with floppy long fur that the explorers had first seen on the trail, and that, with Suydam's support, would become officially recognized by the American Kennel Club in 1935. Although shipping animals overseas was difficult, Suydam sent dogs to the Dalai Lama as well, including pairs of dachshunds and Dalmatians.

The trail had changed who Suydam fundamentally was, and now, instead of pursuing the grandeur of hunting trophy animals, he focused his study on Tibetan Buddhism. "It is a corollary of reincarnation that all life," Suydam explained, "human and animal, is sacred." The explorers had taken animal life, and now, it seemed, they were each finding their own paths to redemption.

After searching for the sacred, Suydam found himself enmeshed in war. On June 4, 1940, Winston Churchill declared, "We shall fight on the beaches, we shall fight on the landing grounds, we shall fight in the fields and in the streets, we shall fight in the hills; we shall never surrender." Suydam was so moved by the parts of the speech reported in the press that he vowed to help the British protect themselves from imminent invasion.

In New York, Suydam started the American Committee for the Defense of British Homes. His group took out advertisements in magazines and spread the message that they were collecting guns for Britain. He asked police departments to donate their confiscated weapons to the cause, as well as appealing to individuals, and, all told, ended up sending more than 3,500 firearms to the UK.

When the United States entered World War II, Suydam served as a colonel in the US Army, just as he had during World War I, although he was now in his fifties. After the war, in his later years, he loved going to the Explorers Club in New York City and astounding audiences with his tales of travel in China and Tibet. He continued his fierce advocacy for endangered species, often citing his work with Kermit saving the Galápagos tortoise from extinction. He loved his Tibetan dogs until the end, breeding Lhasa apsos, or bearded lion dogs, on his farm in New Jersey. He died in 1972 at age eighty-three.

THE WAYWARD SCIENTIST HERBERT STEVENS returned from the trail to a building he knew intimately. The home of the Royal Geographical Society in London had long pulled at his desires, with its vast collection of maps and specimens from around the globe. In 1930, he was returning to the society in triumph, with a vast scientific collection and detailed maps of the Himalayan peaks that would fill in some of the "blank spots" that persisted in the region. Herbert knew that many in the society had doubted that the Roosevelt expedition could be successful, but now they had all been proved wrong. Herbert had finally broken his bad luck streak.

In 1930, the Royal Geographical Society published "Sketches of the Tatsienlu Peaks," Herbert's scientific paper on their expedition, and his first to ever be accepted by the society. His slow pace had paid off—

in August 1929, while the rest of the explorers were already home or headed that way (except for poor Russell Hendee), Herbert had still been on the trail. Unfettered by expectation and knowing that the mountains around Tatsienlu had never been mapped, he'd put his exacting nature to good use. He occasionally set down his butterfly net to assess distances, described glaciers in immense detail, and spoke with locals to assess the respective heights of these uncharted peaks. The maps the society made from his sketches and notes would be a fervent source of pride. He was made a fellow of the Royal Geographical Society and in 1934 published his own account of the trail, *Through Deep Defiles to Tibetan Uplands: The Travels of a Naturalist from the Irrawaddy to the Yangtse.*

Herbert was now in his midfifties, but he wasn't finished exploring. He traveled one more time with Suydam in 1930 to India, where the pair took a slow trail over steep mountains, collecting birds and invertebrates. His last expedition, financed by the zoological department at Harvard University, sent him to Papua New Guinea where he spent a year recording 207 bird species, and uncovering a new subspecies.

Yet for all his travels, the months spent on the trail in China and Tibet were by far his favorite. "Long live the land of the lamas!" Herbert would write. "May no modern inventions ever encroach to disturb the serenity and repose of your land.... Is it also too much to expect, at least one place on God's earth will be left in peace from 'the destructive grasp' of base commercialism?"

Finally recognized by leading scientific societies, Herbert would devote much of his life to sorting through the 12,633 mammals, birds, reptiles, amphibians, insects, and plants the party had brought home from China. He retired in Hertfordshire with his wife, Amy, where he continued to study his vast personal collection of bird eggs and nests, insects, and over four thousand bird skins. "Stevens has done a great deal of excellent work," said Steven Baker, a prominent British ornithologist, "though still too much of it remains unrecorded and locked in his brain." He passed away in 1964 at age eighty-seven, leaving his scientific collections to museums and the vast majority of his money to animal charities.

AS IMPRESSIVE AS HERBERT'S SCIENTIFIC studies were, nowhere was the influence of the trail stronger than in Jack Young. Jack came back to the United States confident, skilled, and roughly $3,000 richer. To him it was an enormous sum, worth some $50,000 in today's money, although a mere fraction of the $15,000 (nearly $300,000 in 2024 dollars) that each of the Roosevelt brothers received. He also returned with a new name. He christened himself Jack Theodore Young, a tribute to Ted and Kermit that he would carry with him for the rest of his life.

Jack had earned the hard-won respect of every naturalist on the expedition team and proved in some part to the curator at the Field Museum in Chicago, known to mistrust Jack for his Chinese heritage, that he'd been wrong to doubt his abilities. He joined the Explorers Club in New York, an exclusive group of adventurers and scientists, where he made new connections and began to dream about what would come next.

Jack applied to the University of Chicago in 1929, hoping to follow in Herbert's footsteps and become the kind of scientist who earned freedom to explore the globe, discovering new species and helping to further human understanding of the natural world. With his experience and glowing recommendation letters, he was speedily accepted into the zoology program in Chicago, but problems soon arose. Even with the money earned from the trail, Jack still needed a job to work his way through school. He applied to the Field Museum for a low-level position, believing that the process would be simple given his credentials. After all, the exhibit halls of that very museum were now filled each day with new panda-obsessed visitors. Instead, he was rejected. The museum would claim that economic pressures from the Great Depression were to blame, although the institute hired other naturalists during this period.

Stuck in New York and frustrated, the only thing that made sense to Jack was to keep exploring. Within the dark, wooden confines of the Explorers Club in New York City, he met Terris Moore, Richard Burdsall, and Arthur Emmons III. Together the group decided to climb Minya

Konka, in the Sichuan Province of Southwest China. From glimpses on the trail and from the maps made by Herbert Stevens, the mountain appeared to be impossibly tall. Herbert's estimated elevation of thirty thousand feet would make it the tallest in the world.

Jack brought his intimate knowledge of the area and expertise as a naturalist, while his three companions offered up their connections and funding from the American Geographical Society. Ted assisted the group by helping them obtain permits, although this time the Chinese government wanted something specific in return. While Western explorers were bringing panda skins to Europe and the United States, China had none of its own exhibits to boast. In return for allowing Jack and his companions to make the trip, they required Jack to shoot animals for Chinese museums. He agreed, and the explorers were off. But just two weeks after they arrived, in late January 1932, the Japanese bombed Shanghai. Jack and the three American men were immediately drawn into fighting rather than exploring, defending the city they happened to find themselves in. For Jack, the intervention of war would become a theme that would define his life.

Although the expedition eventually reached the summit of Minya Konka six months later, they found the elevation to be far less than expected, a mere 24,500 feet. Disappointed by the mountain, Jack now had to fulfill his end of the bargain and start hunting for China. He did so, balancing the scales of the world's museums, at least partially.

In addition to the animals he hunted for China's glory, Jack was able to hunt another panda, this one destined for the American Museum of Natural History. A far cry from the $45,000 that the Field Museum had paid for the Roosevelt specimen, Jack offered his animal for just a hundred dollars. Even this price, however, was difficult to obtain. It took Jack a letter from Kermit, a mention of Suydam, and his own suggestion of a wealthy patron who might pay the fee, in order for the museum to accept. The truth was, the museum was saving its money to buy a specimen off a white explorer, which it would do just a few months later, even though the bear itself was "a headless and footless specimen, difficult to mount."

THE SUMMER WHITE HOUSE

The experience was another harsh lesson for Jack, teaching him how challenging it was to make a living as a naturalist, especially as an Asian American. So many of those he admired, such as the Roosevelts and Suydam Cutting, came to the field with personal wealth and the support of rich donors. Still, he did have friends. Aware that Jack was struggling, Suydam, Ted, and Kermit invited the young naturalist to Oyster Bay, often called the Summer White House during Theodore Roosevelt's tenure. There they began to plot the future, and the Roosevelt brothers and Suydam jointly financed an expedition for Jack in 1934.

It was his first time leading his own expedition. The journey would take a romantic twist when Jack married a Chinese American woman named Adelaide Su-Lin Chen in 1934, and the trip became a combination exploration/honeymoon. Jack had married a remarkable young woman. Su-Lin was born and raised in New York City and had attended college at Wesleyan. She shared her new husband's sense of adventure and love of nature and proved herself fearless on the trail. On that expedition, she became one of the first American women to explore the Himalayas.

As kind as the support from his friends was, Jack's dreams of leading scientific expeditions on behalf of institutions like the Field in Chicago never relented. In 1937, he finally got his wish. He was twenty-eight years old when the American Museum of Natural History offered him $2,700 to head an expedition, collecting plant and small mammal specimens in Tibet and China. It would make Jack the first-ever Asian American leader of an AMNH-funded expedition. After much disappointment and hardship, it was the culmination of everything he'd been working toward the past nine years. The trail had finally paid off.

Ultimately World War II changed Jack's fate, as it did for his fellow explorers. Just like in 1932, Jack's pursuit of a life in science was derailed by war in 1937. He was poised to lead his expedition in Tibet when he came down with appendicitis. While recovering in a hospital in Shanghai, he was suddenly drafted into the Chinese army after the Japanese invaded the country that July. Jack sent his wife to New York for her safety, where his employers at the museum berated Su-Lin for Jack's sudden absence, seemingly oblivious to the impacts of war.

Jack served as a major general in China's army until after the bombing of Pearl Harbor in 1941, when he was drafted into the US Army. He eventually became an intelligence agent, performing missions behind enemy lines in multiple countries across Asia where his cultural intelligence and language skills were essential. He was an American hero: a brigadier general in the American army who earned the Medal of Freedom, two Silver and three Bronze Stars, three Legion of Merit awards, and the Congressional Gold Medal. "I fought for my country," Jack proclaimed, his identity crystallized. Memories of the panda hunt with the Roosevelts would sometimes charm him and at other times haunt him. He died in 2000 at ninety-one years old. His daughter, Jolly Young, would remember her parents fondly as explorers and scientists.

WHILE TED WAS HAPPY TO contribute to Jack's furtherance of science, he had no plans to make any travels of his own. For the first time in years, Ted was determined to settle in New York without the lure of scientific expeditions or political ambitions to pull him elsewhere. He was turning to his first love, books, and in 1932 he edited a book of war poetry, *Taps: Selected Poems of the Great War*, which included verses from Rudyard Kipling, Edith Wharton, Robert Frost, and A. A. Milne, among others.

That book, one of several Ted edited in the early 1930s, led him to what seemed to be his destiny. In 1935 he accepted a job as a publishing executive at Doubleday, Doran. It "has been in the back of my mind . . . for many years," Ted would explain. It was the perfect position for a man who once declared, "I would feel as desolate without a book in my pocket as if I had lost my trousers." The man who loved to read Jane Austen and *Jane Eyre* on the trail had found the ideal job.

Ted brought his fascination with China and the works of Chinese authors with him. He took pride in signing Soong Mei-ling—otherwise known as Madame Chiang, first lady of the republic of China—in 1937 for what he described as the "only book in English written by anyone who is shaping the destinies of the 400,000,000 people who compose China."

Ted continued to write his own books as well. In 1937 Doubleday published *Colonial Policies of the United States*, combining elements of his own life with his meditations on American involvement overseas. The book surprised many for its backing of independence for American territories and descriptions of the dangers of American expansion.

It was as if Ted had finally caught up to his father on that trail in the bamboo forest, and in so doing, freed himself from the weight of expectations. "Ted's truly remarkable career was to be cloaked inevitably and perpetually by the shadow of his father's fame," as his wife, Eleanor, would put it, yet perhaps crawling out from under the heavy burdens of a remarkable career in what his father had been known to do was what Ted had needed all along.

Ted had never owned a house before, so he and Eleanor decided to build one, a large two-story brick home they called Orchard Bay on a slice of his mother's apple orchard, just a stone's throw from the Summer White House. He finally had a permanent space for all the souvenirs he'd collected during his travels, and the rooms filled with knickknacks and treasures and furniture he'd purchased on the trail and in his travels across China. Snug in his home, writing poetry, editing, and reading books that "overflowed into every room in the house, both upstairs and down," Ted frequently worked beneath the gaze of the Wheel of Life, the artwork that the guardian of the eastern border had given him in Muli. His life had come full circle.

Kermit's conclusion was not so satisfying. Alcoholism, extramarital affairs, and disagreements with his brother had left him miserable. His shipping business struggled, and although he was offered new museum-funded expeditions, he refused them all. He had no desire to hunt pandas or, it seemed, any other animal. Yet even amid the ruins of the life he had once led, Kermit kept finding hope. Following his deep regrets after the panda hunt, he dedicated himself to the animals he had once hunted.

In 1930, Kermit took an executive position with the New York Zoological Society. It was a job he badly needed. He and his wife, Belle, were buckling under expenses and cutting costs wherever they could, right down to their newspaper subscriptions and club memberships. The Roosevelts were no strangers to being on the boards of organizations—the positions, often paid, were a perk of being the offspring of a president—but in this case, Kermit had no qualifications beyond the panda hunt. He had earned renown among scientists and conservationists and proved himself capable of more than just a bland corporate board position. For Kermit, this was a job that, despite his mounting personal problems, he cared deeply about.

The zoological society's aims were to advance wildlife conservation, protect endangered species, and promote scientific study. In his role, Kermit helped present data on endangered wildlife with an emphasis on effectively solving conservation problems. He dealt with domestic issues, protecting natural habitats and sanctuaries in the United States, and funded research studies on endangered flora and fauna. He even embarked on new conservation studies intent on preserving species instead of killing them.

In 1930, Kermit and Suydam reunited aboard Vincent Astor's massive 263-foot yacht, the *Nourmahal*. Astor was the heir to a New York real estate fortune worth hundreds of millions of dollars, and frequently entertained the rich and wealthy on his diesel-powered boat, including the president-elect and Ted's nemesis, Franklin D. Roosevelt. Yet in addition to boasting luxury cabins, the yacht was also a "floating laboratory." Thanks to Kermit's influence, the vessel was filled with scientific equipment and clean rooms for studying animals, and was fitted with powerful deep-sea dredges for exploring the sea floor.

It was these features that Kermit and Suydam were most excited for as the yacht motored through the Panama Canal and toward the Galápagos Islands. The islands had been named for the massive numbers of tortoises that once dwelled there, at one time more than two hundred thousand, comprising an expansive fifteen species, but their numbers were in sharp decline. Whalers had been hauling the animals away for years, often two to three hundred at a time, usually taking the females since they were bigger. The largest of the tortoises spanned five feet and weighed more than five hundred pounds. The animals were a source of fresh meat that could be easily stored on board without spoiling, as tortoises can live for up to 177 years. With a preference for kerosene growing in the 1850s, the whaling industry collapsed, and sadly, it managed to take down the giant tortoise with it.

"This animal," wrote Kermit, "formerly of great food value, is threatened with extermination in the only region it ever existed." Kermit and Suydam arrived in the Galápagos to explore the islands, note habitat destruction, and collect unusual plant species at a leisurely pace that would have pleased even Herbert. They then moved on to the tortoises, identifying different species on separate islands and noting their size, diet, and habitat. It was work they were made for.

The next part was more difficult. Kermit and the zoological society had determined that a breeding program in the United States was the only way to save the endangered animal. Making use of Astor's beautiful yacht, they brought tortoises of both sexes aboard and prepared to make them comfortable for the long trip home. For Kermit, the experience was unique. He was not bringing home shells or skins to hang in his drawing room, or specimens to be picked apart in a museum and eventually stuffed in an exhibit. He wasn't even bringing home live animals to dwell in a zoo. Instead, he was trying to *save* a species according to a carefully designed breeding program. It would take decades, but the program ultimately proved successful, with the *New York Times* praising their work in "the Galápagos Islands where they helped save from extinction the giant land tortoise."

In 1935, Kermit became president of the Audubon Society, amid the environmental crisis that defined the Dust Bowl in the American Midwest, while keeping his job with the New York Zoological Society.

As head of the Audubon Society, he wrote a monthly column called "The President's Page" for its publication, *Bird-Lore*. From the beginning, his writing was focused on environmental conservation and encompassed strong praise for unappreciated species of plants and animals. In his columns he argued for the preservation of the bison, and for the protection of a grove of Sitka spruce near the entrance to Yellowstone National Park. The addition of a visible and activist face was a boon to the organization, and Kermit was not shy about making his mark. Although he knew it was a move that would enrage thousands, he boldly called for a national closure of duck hunting his first year. The nasty letters began to stuff his mailbox, but Kermit believed the move was essential, and managed to convince many leaders of hunting organizations to follow suit.

It still wasn't enough, as Kermit kept pushing ahead. In his role at the zoological society, Kermit began reaching out to leaders of foreign nations, particularly China, for support in expanding the cause of conservation. In correspondence with representatives from China, Kermit emphasized his fear for the panda's future. "All hunting must cease," Kermit wrote of the panda in 1936.

It was a bold statement for a man who had not only personally profited from hunting it but then sparked an entire industry focused on the panda in the United States. Between 1936 and 1946, fifteen pandas were stolen out of the Chinese wilderness. Depending on whether they were dead or alive, they went to American zoos or natural history museums, drawing hundreds of thousands of visitors.

Unlike most of the explorers and hunters tracing his footsteps in Southwest China, Kermit knew from experience how elusive the black-and-white bears truly were. He remembered how the Yi hunters had seldom seen the animals. While in the United States the sweet face of the black-and-white bear was already plastered on sparkly brooches and straw-stuffed toys, the animal was suspiciously absent in the history of Chinese art, on display in numerous exhibits in New York in the 1930s. These artworks portrayed a wide range of animals, from fish to cranes to deer, even including mythical creatures such as the phoenix and colorful dragons, but never the giant panda.

Kermit's passionate correspondence with Chinese diplomats made an impression. In his letters he discussed the large number of pandas that were "dying in captivity," whether during the long voyage west, or in zoos in the United States. "The only solution," Kermit concluded, "is a complete ban."

He wasn't alone in pushing for these changes. Even Ruth Harkness, the socialite who'd brought the first living panda to New York, had experienced a change of heart. In 1938 she underwent what she described as an "arduous journey" to return the last of the three panda cubs she'd captured over the years, and the only one to survive, back to the bamboo forests.

In 1938, asks of this nature were far outside the mainstream. Hunting bans were rare, even for animals that were on the brink of extinction. Such was the case for sea otters, which once spanned the entire Pacific Rim, from the islands of Japan across to the West Coast of the United States and down into Baja, Mexico. The marine mammal, whose fur has more hair per square inch than any other animal, was a victim of the fur trade, and by the late 1800s the population had been decimated. An international treaty signed in 1911 banned hunting of the sea otter, but the law didn't apply to the California coast where hunting was still legal according to state law. Since sea otter pelts fetched an enormous sum, more than $1,500 each, the situation seemed hopeless. By the time state legislators banned hunting them in 1913, the species was already considered extinct.

Then, out of nowhere, a raft of sea otters was spotted off the coast of Big Sur, California, in 1937. *Life* magazine printed a picture of the animals over the headline "The 'Extinct' Sea Otter Swims Back to Life." It would take the species decades to reclaim a fraction of its former territory, turning the sea otter into a near mythical animal, seen by so few humans over multiple generations that it was as if it had vanished entirely.

This was the future that awaited China if it didn't take immediate action. The panda, already on the cusp, was poised to disappear entirely. In 1938, China banned all giant panda hunting in the country, the first trophy game animal to receive such protections. The government also cracked down on exporting live animals, implementing a stringent permitting process that prevented helpless panda cubs from being sent off to lead short, erratic lives on foreign soil. That same year the panda earned its place as a symbol of China and became emblematic of conservation efforts across the globe.

In 1941, Madame Chiang—Soong Mei-Ling, the author Ted had published at Doubleday a few years earlier—reopened the doors of panda diplomacy, giving the Bronx Zoo a pair of the highly valued cubs as a symbol of China's gratitude for America's support of Chinese military efforts against Japan. Madame Chiang announced that she hoped "their playful antics will bring as much joy to American children as American friendship has brought to our Chinese people."

Just like the stolen animals that preceded them, the cubs wouldn't live long. The species was still newly discovered, and zookeepers had no experience in taking care of pandas. As scientists would learn in the decades ahead, a thorough biological understanding is necessary to foster sensitive and endangered species, and even that, many times, is not enough. By the 1950s, there were no longer living pandas in the United States. Every single one had died. The only places left to see the animals were museum exhibits where their preserved, dried forms persisted among simulated white snow and crunchy fake bamboo. It would stay that way until 1972.

Famously fragile in captivity, known to breed infrequently, and pleased to live out of view of those animals—us—who've brought it only trouble, the panda will ultimately come and go. Sometimes the living creatures will light up zoos the world over like the sun rising over the Himalayas, setting the snowy shoulders of the mountains aflame. Other times, in moments of peril or scarcity, they will persist in the shadows. We would do well to appreciate the knowledge that somewhere in the wilderness, not in a zoo nor in a breeding facility, but rather in the dense bamboo-thick wilds of the Himalayan mountains, there are wild giant pandas in near-silent contemplation living just beneath the clouds.

The knowing will have to be enough.

EPILOGUE

TRAIL'S END

Fate is not always kind to the sons of famous men, but toward Kermit Roosevelt it was particularly cruel. The reality of Kermit's life was starkly different from the impression the world had of him. He was often described as quiet and inquisitive—not boisterous like his father, nor cheerful like his elder brother, but introspective and sometimes depressed. His work in nature conservation during the decade following the panda hunt was a stunning act of redemption, but by the fall of 1939, his thoughts turned to what was happening in Europe. When Germany invaded Poland, Kermit knew he had to join the fight. Although he could have taken a position plotting strategy and orchestrating shipping logistics, he instead chose to enlist as a regular commission in the British Army, just as he had done as a much younger man during World War I.

Although trained as a machine gun specialist, he was no longer the resilient twenty-nine-year-old who could rush off into battle. At age fifty, the years of sadness and alcoholism had weakened him, both mentally and physically. To treat his alcoholism, he had been given the drug paraldehyde, a sedative used at the time to treat alcohol withdrawal. Instead of benefiting from it, Kermit became addicted to the colorless liquid, which breaks down into toxic by-products, leading to liver disease. Addiction and mental illness had become tightly intertwined in his body, like a strangler fig enveloping its host with increasing pressure, poised to kill.

Still, he persevered, believing strongly that the Germans had to be defeated. By 1940, he had been made a colonel in the British Army and was leading a force to aid Finland during the country's winter war with the Soviet Union. The mission was aborted; before their group of volunteers arrived, the nation surrendered to their invaders. Kermit was sent to Norway instead, fighting alongside the British in the mountains surrounding the town of Narvik. The operation was not a success, and Britain was forced to retreat, but Kermit was noted for "helping get men and equipment out and even carrying some of the wounded on his back."

Next the British sent him to Egypt, where the pace was slower, and Kermit began drinking again. His health deteriorated so quickly that the army had no choice but to send him home. Kermit was devastated and, although badly off, he convinced his friend Winston Churchill to advocate on his behalf. "Major Kermit Roosevelt has been to me in great distress," the prime minister wrote his secretary of state for war. "Doctors have marked him 'E' this morning, thus putting him out of the Army. When he came originally to me at the Admiralty in October 1939, I considered it a matter of political consequence that his wish to serve with us in the fight should be granted. I thought it symbolic. I still think that he should not be treated as an ordinary case, and if he wishes to go on with us he should be allowed to do so. Will you very kindly look into the matter? His morale is very high. And he is very unhappy at the idea of being invalided out now."

Not even Churchill's influence could save him. Kermit was forced to resign his commission on May 5, 1941. When the United States entered the war on December 7, however, Kermit was quick to join up. His health hadn't improved—if anything, it had deteriorated, thanks to his continued drinking—but with the help of his cousin FDR, Kermit received a commission in the US Army at Fort Richardson, Alaska, albeit far removed from the military action he craved.

TRAIL'S END

"There is a universal saying to the effect that it is when men are off in the wilds," wrote Kermit, inspired by Jane Austen's lively language, "that they show themselves as they really are." There had been short trips here and there, but Kermit had last been himself in 1929. That was the final time he would ever be immersed in the wilderness, exploring alongside his brother. The panda hunt had forever altered his life.

Fort Richardson was experiencing an unusually balmy Alaska summer in 1943. On the evening of June 3, Kermit turned to a friend and asked what he was planning to do when they returned to the barracks. "Sleep," his buddy replied.

"I wish *I* could sleep," Kermit said. That night he took out an old revolver, placed it under his chin, and pulled the trigger.

When Ted heard the news, he remarked sadly, "He really died five years ago." Perhaps Ted could have traced it back even further, to 1929, when the panda fell in the forest and life began to fall apart. A dark shadow had fallen across their lives the moment the two brothers had simultaneously pulled their triggers. The passing years had been defined by acts of redemption and moments of joy, but Tibet and China had fundamentally changed Ted and Kermit, and there could be no going back to the way things were.

Ted would live only a year longer than Kermit. Like his younger brother, he joined the war effort early, in 1940, and despite serious health issues and a weak heart, he was placed on active duty at his earnest request in 1941. Ted led numerous operations overseas, including in North Africa and Italy. In 1944, he was sent to England to assist the Normandy invasion, where he wrote three successive pleas to be allowed to accompany the troops landing ashore on D-Day. Although he was fifty-six and walking with a cane, his request was approved, with mixed feelings. Some military commanders "felt sure he would be killed."

Ted carried his cane as he landed on the beaches of Normandy. He was the oldest soldier in the first wave of assault troops, and the only general. After reconnaissance and plotting their attack positions, Ted yelled from Utah Beach, "We'll start the war from right here!" Many soldiers were inspired by Ted's presence on D-Day, and although 2,501 Americans died on June 6, 1944, Ted was not one of them.

Thirty-six days later, on July 12, 1944, the weak heart that Ted had been hiding from his military doctors caught up with him. In France, Ted woke in the middle of the night with chest pain. He died of a heart attack at midnight.

The explorers are gone, but so too are many of the species they encountered along the trail. The Chinese paddlefish, which could once grow to more than twenty-three feet in length, went extinct in the early 2000s, likely the result of overfishing and dam construction. It may soon be joined by the "Crim," the river monster turned dolphin, a species now critically endangered in China. Industrial development has destroyed the deep pools and complex habitat that once allowed the animals to thrive.

So many of the bird species the explorers collected are absent from the Chinese and Tibetan wilderness today. The sarus crane that Herbert loved with a passion and would not allow anyone to shoot is endangered. Herbert described them as grouped in large clusters on the edges of rice paddies, but now they are rarely seen. The draining of wetlands has decimated their once abundant habitat. The colorful hornbills that Kermit admired on the trail have also mostly disappeared. The species is critically endangered and found in such small populations that scientists worry they may be gone for good.

While nineteen new species were described by the 1929 expedition, one in particular was special to the Roosevelts. *Muntiacus rooseveltorum,* or Roosevelt's barking deer, was the only species named after the brothers. It wasn't a name they chose themselves; all scientific designations were given by the taxonomists studying the species at the museum. This deer, however, was named after Ted and Kermit as a tribute to their efforts on the trail.

Roosevelt's barking deer is small, covered in reddish-brown fur, and found only in remote mountains. Yet nearly as soon as the species was discovered, it seemed to disappear. The explorers found the animal in

1929—and then that was it. Trail camera footage in Vietnam in 2014 might have spotted a glimpse of the species, but no one can be certain.

The truth is that species are disappearing faster than scientists can document them. This is particularly evident in Southwest China where animal diversity is incredibly high. "It's often easier to find animals from 10 different species," says Dr. Han Lianxian, a zoologist at Southwest Forestry University, "than find 10 different individuals of the same species. Each species only has a small population and is very fragile." A 2021 expedition along the Mekong River uncovered 175 new species, while one that took place in 2022 discovered 205 new species of plants, amphibians, fishes, and even a new mammal. The forests are a treasure box, one whose contents are being raided before we can even document what we have.

One bright spot is the animal that the Roosevelts once craved above any other: the giant panda. While we might look at the 1930s as a time of barbarism, pandas have suffered far more in our modern history. Intense environmental destruction, along with poaching in the 1980s, nearly wiped out the species entirely. "I can only view with irony," writes biologist George Schaller, who researched the giant panda in the 1980s, "the fact that never has the panda's destruction been as rapid as during the years we studied it, during a period when it received more attention than at any time in its long history."

Since Schaller described the thorny nature of conservation politics in the 1980s, the situation has improved. Multiple laws enacted by China have aided intense conservation efforts, including bans on deforestation and other habitat protections. Nearly two-thirds of all wild pandas are protected on sixty-seven panda reserves in thick, mountainous bamboo forests.

After decades of uncertainty, in 2021 the giant panda was moved from endangered to vulnerable on the red list of threatened species. From its low in the 1980s, when there were as few as 1,114 pandas living in Asia, the species has made a comeback. Still, there are fewer than two thousand pandas in the wild, making them one of the rarest mammals on earth.

"It is the difficult fate of this generation," writes George Schaller in his book *The Last Panda*, "to finally grasp the magnitude of all the offenses against the panda and other forms of life; the extent of environmental destruction has been nothing less than a spiritual divestment, a renunciation of past and future."

Habitat protections put in place for the panda have benefited other animals such as the takin, the unusual Bovidae species the Roosevelt brothers met on the trail that can reach up to 770 pounds but is rarely seen in nature. Because the animal occupies the same bamboo forests as the panda, its population has remained largely intact.

Other species have not been so fortunate. Although the panda is considered to act as both a "flagship" and an "umbrella" species, protecting other threatened animals in its vicinity and even around the globe, that has not always been the case. Due to the combined effects of poachers, deforestation, and disease, a 2020 study found, leopards have disappeared from 81 percent of giant panda preserves, wolves from 77 percent, and dholes, an Asiatic wild dog, from 95 percent. It is a profound loss. In danger are also the golden monkeys. When the Roosevelts and their guides shot the animals outside of Muping, they couldn't know how quickly the species would become endangered, with fears that because of logging, their canopy community may never fully rebound.

For the panda itself, it is nearly as difficult to find one in the wild today as it was a hundred years ago. One scientific study assessing panda population dynamics stated that "a survey based on direct sightings is completely unrealistic for this species due to the extreme difficulty locating pandas." The black-and-white bears are more easily found at the Chengdu Research Base and Wolong Panda Center, two preserves that integrate scientific research and captive breeding.

Science is like a city skyline, always reaching for more. Each research study builds on the last, creating a chain of human understanding. Where once hunting large mammals was seen as an irreplaceable instrument of science, it is now rare. With the last of the large mammals categorized, biologists are more likely to carry a tranquilizer gun than a shotgun, and their trophies are vials of blood and tags of skin rather than stretched and dried furs.

To our modern eyes, the shooting of the panda by the Roosevelts seems cruel and unnecessary. Yet how will future generations assess our actions today? Soil erosion, pollution, deforestation, and the multitude of human-driven causes of climate change may be viewed as even more brutal and unforgivable than shooting an innocent animal in the woods. Like the Roosevelts in 1929, we do not yet know what lasting impacts and unintended consequences our actions will have.

What we can be certain of is that safeguarding the giant panda has expanded the protections for thousands of species beyond Tibetan and Chinese borders. The panda, as a symbol of conservation, has influenced ecological conservation efforts around the globe. Thanks to its example, new protections have been given to elephants in Asia and Africa, twenty thousand of whom are slaughtered annually as part of the ivory market. Efforts are also underway to protect the world's tigers, with conservation programs expanding the protection of their massive territories in an effort to double their numbers across the globe.

Among the peaks of the Himalayas there are new scientific expeditions forming. The naturalists come from across the globe. They take planes, buses, and cars to their destinations; no longer are they expected to walk a thousand miles on foot, yet many will still follow the fragmented pieces of a vanishing trail. In their hands, amid the bounty of technology, remain slim paper field notebooks. Numbers will be recorded on those pages, handwritten in ink and sometimes pencil, that document a world few of us will ever glimpse. Under their watchful eyes, and often still requiring death, new species will be uncovered, described, and named.

Even at the end of the trail the journey is not over. Instead, the field journals build up over the decades, an ever-expanding record of a disappearing world.

ACKNOWLEDGMENTS

Thank you, readers—this book would not be possible without you. I am so grateful for those who have written me letters, come to my author events, and contacted me online.

Special thanks to my grandfather, the late Sheldon Katz, who lived in China for more than a decade and who was kind enough to take me on adventures and ignite my love for the country, its wilderness, its languages, and especially its people.

I'm eternally thankful for Michael Kiefer, who graciously let me comb through his interviews with Tai Jack Young and shared his immense expertise in the history of the panda. I'm very grateful for the materials he shared with me, which gave me greater insight into Tai Jack Young's remarkable life and inspirations. Similarly, I'm fortunate to have spoken with Jolly King, who shared memories of her family.

My editor, Nick Ciani, not only took a chance on this book but worked his magic on the text. This book would not be as good as it is without his masterful edits and needed cuts. I'm very grateful for the entire team at Atria/One Signal, including Abby Mohr, Mark LaFlaur, and Hannah Frankel, among many others, who have given so much of their time

ACKNOWLEDGMENTS

and expertise to this project. Every writer hopes to have someone like Laurie Abkemeier on their side as an agent and editor, and I have been fortunate to work with her on all my books. Her enthusiasm and dedication are an inspiration to me.

This book would not have been possible without the help of the many archivists and historians who assisted me in finding images and primary sources from Ted and Kermit Roosevelt and their famous father. I'm particularly indebted to Rebecca Wilkie at the Field Museum of Natural History, Mai Reitmeyer at the American Museum of Natural History, and Courtney Matthews at the Library of Congress. I'm also grateful to the scientists who generously spent time with me, going over the biology and botany of the manuscript. Special thanks go to Dr. Tim Caro, Dr. Lei Zhang, and Dr. Hua Zhu.

Graham Snow (@mapograham) is an incredible artist and mapmaker. The map he created for this book adds so much to the narrative, and I'm very grateful for his hard work and exceptional abilities.

For my friends: bestie Darcie Jarrett Tuite, Mark Tuite, Rebecca and Evan Nuckles, Angelee Brockmeyer, Anna Seltzer, Kelly Nguyen, and my BRF Brittany Brubaker. For my Met Hill fam: Rachael and Gerry Coakley, Susie and Ben Bird, Elizabeth Keane, Sean Cashman, and Sarah Elliott. My Ventucky crew: Emlyn Jones, Elizabeth Shaw, Amy Cantor, Scott Ambruster, Jennifer and Payson Thompson, Tim Flanagan, J. A. and Joline MacFarland. For my little Turkeys—I love you all. For my family: Joyce Boone, Claire McCleery, Rose Grundgeiger, Harry and Dottie Goode, Shane, Frannie, Ruby, Harrison, and Andrew Vesely, Shelley and Dave Buttgen, Scott, Tricia, and Shea Holt.

For those I owe everything to: Larkin Holt, who doesn't believe in book dedications but deserves all of mine, and my dazzling daughters Eleanor Frances Holt and Philippa Jane Holt. Special thanks to the best pup ever, Winnie.

NOTES

ABBREVIATIONS

AMNH	Archives of the American Museum of Natural History, New York
FM	Archives of the Field Museum, Chicago
KBRP	Kermit and Belle Roosevelt Papers, Library of Congress Manuscript Division
LOC	Library of Congress
MKC	Michael Kiefer Collection (private papers and interview notes of author Michael Kiefer)
NYPL	New York Public Library
TRC	Theodore Roosevelt Collection, Harvard Library
TRP	Theodore Roosevelt Papers, Presidential Papers Series, Library of Congress Manuscript Division
TRJP	Theodore Roosevelt Jr. Papers, Library of Congress Manuscript Division

NOTES

PROLOGUE: THE LAST LARGE MAMMAL

xii *"We did not let even":* Theodore Roosevelt and Kermit Roosevelt, *Trailing the Giant Panda* (New York: Charles Scribner's Sons, 1929).

xiii *In Egypt, King Ptolemy II had a polar bear:* Kieran Mulvaney, *The Great White Bear: A Natural and Unnatural History of the Polar Bear* (New York: Houghton Mifflin Harcourt, 2011).

xiii *called it "spotted bear":* The mythical nature of the panda and the many names that refer to it are explained in Elena Songster, *Panda Nation: The Construction and Conservation of China's Modern Icon* (New York: Oxford University Press, 2018).

xiii *"While there are tantalizing stories":* Tim Brady, *His Father's Son: The Life of General Ted Roosevelt Jr.* (New York: Penguin, 2017).

xiii *While stationed in the remote mountains:* Foster Stockwell, *Westerners in China: A History of Exploration and Trade, Ancient Times Through the Present* (Jefferson, NC: McFarland, 2015).

xiv *Some researchers trace its roots:* Richard Perry, *The World of the Giant Panda* (New York: Taplinger, 1969).

xv *"A giant panda," read the announcement: Natural History* 19 (October–May 1919), AMNH.

xv *Ted and Kermit's connections with the museum:* Edmund Morris, *The Rise of Theodore Roosevelt* (London: Collins, 1979).

xvi *The elephant calf he shot:* Kermit Roosevelt, *The Happy Hunting Grounds* (New York: Charles Scribner's Sons, 1920).

xvi *"I can be condemned":* Patricia O'Toole, *When Trumpets Call: Theodore Roosevelt After the White House* (New York: Simon & Schuster, 2005).

xvi *Naturalists of the era:* Henry Fairfield Osborn and Harold Elmer Anthony, "Can We Save the Mammals?" *American Museum Journal* 22, no. 5 (September–October 1922).

xvi *In Uruguay in 1833:* Adrian Lister, *Darwin's Fossils: The Collection That Shaped the Theory of Evolution* (Washington, DC: Smithsonian Books, 2018).

xvii *"Should some interesting mammal":* Osborn and Anthony, "Can We Save the Mammals?"

xvii *Stocks had quadrupled in value:* John Kenneth Galbraith, *The Great Crash, 1929* (New York: Houghton Mifflin, 1961).

NOTES

xviii   *was worth some $400 billion:* Ron Chernow, *Titan: The Life of John D. Rockefeller, Sr.* (New York: Random House, 1998).

xviii   *was called upon, not once but twice:* Jean Strouse, *Morgan: American Financier* (New York: Random House, 2012).

xviii   *Wealth inequality today is peaking:* Matthew Stewart, *The 9.9 Percent: The New Aristocracy That Is Entrenching Inequality and Warping Our Culture* (New York: Simon & Schuster, 2021).

xviii   *"Neither Kermit nor I can afford this":* T. Roosevelt and K. Roosevelt, *Trailing the Giant Panda*.

xviii   *Their father, Theodore Roosevelt, had inherited $60,000:* "Will of Theodore Roosevelt," *New York Times,* February 17, 1878.

xviii   *Roosevelt lost most of his fortune:* David McCullough, *Mornings on Horseback: The Story of an Extraordinary Family, a Vanished Way of Life, and the Unique Child Who Became Theodore Roosevelt* (New York: Simon & Schuster, 2007).

xviii   *The land and most of his fortune were left to his wife, Edith:* Theodore Roosevelt, "Last Will and Testament of Theodore Roosevelt," TRC.

xviii   *In 1925, the Field Museum had funded:* Theodore Roosevelt and Kermit Roosevelt, *East of the Sun and West of the Moon* (New York: Charles Scribner's Sons, 1926).

xix   *were just the thing to tempt hunters:* Brian Herne, *White Hunters: The Golden Age of African Safaris* (New York: Henry Holt, 1999).

xix   *"Central Asia . . . is the mecca of our desires":* T. Roosevelt and K. Roosevelt, *East of the Sun and West of the Moon*.

xix   *"It's a warning":* Kermit Roosevelt to John Winant, 1928, KBRP.

CHAPTER 1: THE HAPPY VALLEY

2   *Tai Jack Young looked down:* Michael Kiefer interview with Tai Jack Young, January 12, 1990, MKC.

3   *His last name had already been changed:* Michael Kiefer, *Chasing the Panda: How an Unlikely Pair of Adventurers Won the Race to Capture the Mythical "White Bear"* (New York: Four Walls Eight Windows, 2002).

3   *thanks to his upbringing in a small village outside of Hong Kong:* Ibid.

NOTES

3 *reflecting his own fractured identity:* Michael Kiefer interview with Tai Jack Young, January 12, 1990, MKC.

4 *"I am eighteen"; "Well, Colonel"; "I would like to go":* Theodore Roosevelt and Kermit Roosevelt, *Trailing the Giant Panda* (New York: Charles Scribner's Sons, 1929).

4 *"He was a slight, nice-looking boy":* Ibid.

4 *"Doubt if Chinaman":* Michael Kiefer, "Quentin Young, Ruth Harkness, and the Pandas in China," *San Diego Reader*, November 29, 1990.

4 *"We have only an hour":* T. Roosevelt and K. Roosevelt, *Trailing the Giant Panda*.

5 *Herbert Stevens was a biologist originally from:* Herbert Stevens, *Through Deep Defiles to Tibetan Uplands: The Travels of a Naturalist from the Irrawaddy to the Yangtse* (London: H. F. G. Witherby, 1934).

6 *"This isn't the Taping River":* T. Roosevelt and K. Roosevelt, *Trailing the Giant Panda*.

7 *"Herbert!" Kermit screamed; "Climb aboard"; "My name is Saw Bwa Fang Tao":* Ibid.

8 *"He's inviting us to breakfast tomorrow morning"; "We barely slept":* Ibid.

8 *"If it wasn't for him, who knows where Herbert would be":* Suydam Cutting, *The Fire Ox and Other Years* (New York: Charles Scribner's Sons, 1940).

9 *"He is a man of undaunted courage":* T. Roosevelt and K. Roosevelt, *Trailing the Giant Panda*.

10 *"By the way":* Cutting, *The Fire Ox and Other Years*.

11 *"They were everything to me":* Michael Kiefer interview with Tai Jack Young, January 12, 1990, MKC.

11 *Only in the forest did he feel like himself:* Kermit Roosevelt, *The Long Trail* (New York: Charles Scribner's Sons, 1921).

11 *volunteered among the first soldiers to go to the Western Front:* Tim Brady, *His Father's Son: The Life of General Ted Roosevelt Jr.* (New York: Penguin, 2017).

11 *Ted had leased oil fields on public lands:* Ibid.

11 *"politically obliterated":* Eleanor B. Roosevelt, *Day Before Yesterday* (Garden City, NY: Doubleday, 1959).

12 *"Market square"*: T. Roosevelt and K. Roosevelt, *Trailing the Giant Panda*.
12 *"Backs or mules"*; *"Is your brother thinking of improving the roads?"*: Ibid.
13 *"That's where we'll have breakfast"*; *"Is he here?"*; *"Why spend so much money"*: Ibid.
14 *"Where's your brother?"*; *"Have you heard of the poppy?"*: Ibid.
14 *they would see opium addiction everywhere in China*: Stephen R. Platt, *Imperial Twilight: The Opium War and the End of China's Last Golden Age* (New York: Vintage Books, 2018).
15 *"Those poor souls"*: T. Roosevelt and K. Roosevelt, *Trailing the Giant Panda*.
15 *Thoroughly dominated by the British, China was powerless*: Platt, *Imperial Twilight*.
15 *"Never before or since"*: United Nations Office on Drugs and Crime, "A Century of International Drug Control," *United Nations World Drug Report*, 2008.
16 *"They put her in a cage"*: T. Roosevelt and K. Roosevelt, *Trailing the Giant Panda*.
16 *When Jack was four*: Michael Kiefer interview with Tai Jack Young, January 12, 1990, MKC.
16 *"Will you pick an English name for him?"*: T. Roosevelt and K. Roosevelt, *Trailing the Giant Panda*.
17 *"Walter," Fang repeated*: Ibid.
18 *"Put your guns down!"*: Stevens, *Through Deep Defiles*.
18 *"Antigone antigone"*; *"The terraced slopes of rice cultivation"*: Ibid.

CHAPTER 2: THE VALLEY OF DEATH

19 *Kermit held the .22-gauge*: Theodore Roosevelt and Kermit Roosevelt, *Trailing the Giant Panda* (New York: Charles Scribner's Sons, 1929).
20 *river otters are an indicator species*: D. Gallant et al., "Habitat Selection by River Otters (*Lontra canadensis*) Under Contrasting Land-Use Regimes," *Canadian Journal of Zoology* 87, no. 5 (2009).

20 *river otters were disappearing fast:* Theresa L. Goedeke et al., "Otters as Actors: Scientific Controversy, Dynamism of Networks, and the Implications of Power in Ecological Restoration," *Social Studies of Science* 38, no. 1 (2008).

20 *Their thick, waterproof fur:* Todd McLeish, *Return of the Sea Otter: The Story of the Animal That Evaded Extinction on the Pacific Coast* (Seattle: Sasquatch Books, 2018).

21 *Like his father, he wasn't a particularly good shot:* Kermit Roosevelt, *The Happy Hunting Grounds* (New York: Charles Scribner's Sons, 1920).

21 *"Perseverance, skill in tracking"; "When Father went off into the wilds":* Ibid.

21 *"You don't want to go that way":* T. Roosevelt and K. Roosevelt, *Trailing the Giant Panda*.

22 *In 1913, Kermit and his father had traveled through South America:* Candice Millard, *The River of Doubt: Theodore Roosevelt's Darkest Journey* (New York: Doubleday, 2005).

22 *"The Brazilian wilderness stole ten years of my life":* Ibid.

22 *"It's too oppressive":* Suydam Cutting, *The Fire Ox and Other Years* (New York: Charles Scribner's Sons, 1940).

22 *"You can't imagine what power these women have":* Theodore Roosevelt and Kermit Roosevelt, *East of the Sun and West of the Moon* (New York: Charles Scribner's Sons, 1926).

23 *Such guides were sometimes called porters, and other times coolies:* Ashutosh Kumar, *Coolies of the Empire: Indentured Indians in the Sugar Colonies, 1830–1920* (Cambridge, UK: Cambridge University Press, 2017).

23 *"The entire Chinese coolie class":* Theodore Roosevelt, *A Square Deal* (New Jersey: Allendale Press, 1906).

23 *"More than half our porters were girls and women":* T. Roosevelt and K. Roosevelt, *Trailing the Giant Panda*.

24 *also carried a "stout stick":* Cutting, *The Fire Ox and Other Years*.

24 *"the land of women's emancipation":* T. Roosevelt and K. Roosevelt, *Trailing the Giant Panda*.

25 *"He's a nuisance":* Kermit Roosevelt journal entry, 1928, KBRP.

25 *He could tell the butterfly was a swallowtail:* Herbert Stevens,

Through Deep Defiles to Tibetan Uplands: The Travels of a Naturalist from the Irrawaddy to the Yangtse (London: H. F. G. Witherby, 1934).

25 *the green was the result of an unusual three-dimensional structure:* László Péter Biró et al., "Living Photonic Crystals: Butterfly Scales—Nanostructure and Optical Properties," *Materials Science and Engineering: C* 27, no. 5 (2007).

26 *"Wherever private enterprise has safe-guarded its interests":* Herbert Stevens, "Notes on the Birds of the Sikkim Himalayas," *Journal of the Bombay Natural History Society* 29, no. 3 (1923).

26 *Because so many animals depend on insect populations:* Oliver Milman, *The Insect Crisis: The Fall of the Tiny Empires That Run the World* (New York: W. W. Norton, 2022).

26 *These scientists called themselves "heroic entomologists":* Alvah Peterson, "Are Economic Entomologists Becoming 'Insecticide Minded'?" *Journal of the New York Entomological Society* 40, no. 2 (1932).

26 *On an expedition to Vietnam in 1924:* Amberley Moore, "Herbert Stevens (1877–1964): Collector, Benefactor and Enigma," *British Ornithologists' Club Newsletter,* February 1923.

28 *If they did not act, the women knew they'd be dead:* T. Roosevelt and K. Roosevelt, *Trailing the Giant Panda.*

28 *China was in the midst of civil war:* Diana Lary, *China's Civil War: A Social History, 1945–1949* (Cambridge, UK: Cambridge University Press, 2015).

CHAPTER 3: THE CRIM'S POOL

31 *"sense of oppression":* Theodore Roosevelt and Kermit Roosevelt, *Trailing the Giant Panda* (New York: Charles Scribner's Sons, 1929).

31 *"Neither Kermit nor I really enjoy fully exploring or hunting unless we are together":* Ted Roosevelt to Seymour Halpern, 1929, TRJP.

31 *"Something needs to be done":* T. Roosevelt and K. Roosevelt, *Trailing the Giant Panda.*

32 *He tried every dialect of Chinese he knew:* Michael Kiefer interview with Tai Jack Young, January 12, 1990, MKC.

NOTES

32 *Jack felt himself caught between worlds:* Ibid.
32 *Jack felt adrift:* Michael Kiefer, *Chasing the Panda: How an Unlikely Pair of Adventurers Won the Race to Capture the Mythical "White Bear"* (New York: Four Walls Eight Windows, 2002).
34 *"He's asking what you're looking for":* T. Roosevelt and K. Roosevelt, *Trailing the Giant Panda*.
34 *"Ask for the black-and-white bear":* Ibid.
35 *Kermit knew it must be* Ailurus fulgens . . . *the red panda:* Angela Glatston, *Red Panda: Biology and Conservation of the First Panda* (London: Elsevier Science, 2010).
35 *The species had been described a hundred years earlier:* William Coleman, *Georges Cuvier, Zoologist: A Study in the History of Evolution Theory* (Cambridge, MA: Harvard University Press, 1964).
36 NO FEET BINDING! *it proclaimed:* T. Roosevelt and K. Roosevelt, *Trailing the Giant Panda*.
36 *Although foot-binding had been formally banned:* Wang Ping, *Aching for Beauty: Footbinding in China* (Minneapolis: University of Minnesota Press, 2002).
36 *It's believed that foot-binding was popularized in the tenth century:* Ibid.
37 *a "foolish and barbarous custom":* T. Roosevelt and K. Roosevelt, *Trailing the Giant Panda*.
37 *"a gruesome monster":* Ibid.
38 *"one of the most desolate places of the earth":* Ibid.

CHAPTER 4: EAVES OF THE WORLD

39 *"He's slow":* Suydam Cutting, *The Fire Ox and Other Years* (New York: Charles Scribner's Sons, 1940).
40 *writing the word* Brackish: Herbert Stevens, *Through Deep Defiles to Tibetan Uplands: The Travels of a Naturalist from the Irrawaddy to the Yangtse* (London: H. F. G. Witherby, 1934).
40 *"In the seas":* Pliny the Elder, *Natural History, Volume IX* (Cambridge, MA: Harvard University Press, 1952).
40 *Their raw brain size is slightly larger than that of humans:* Maddalena

Bearzi and Craig B. Stanford, *Beautiful Minds* (Cambridge, MA: Harvard University Press, 2009).

40 *Over the course of history, the human brain has tripled in size:* Yuval Noah Harari, *Sapiens: A Brief History of Humankind* (New York: HarperCollins, 2015).

41 *As earth became increasingly unstable, the evolution of the human brain:* Ibid.

41 *They were called a euryhaline species:* Manuel Ruiz-Garcia and Joseph Mark Shostell, *Biology, Evolution, and Conservation of River Dolphins Within South America and Asia* (Hauppauge, NY: Nova Science Publishers, 2010).

41 *Locals knew the Irrawaddy to be shy:* Ibid.

43 *It had thick, white plumage at the top of its head:* Bikram Grewal et al., *A Photographic Guide to the Birds of India: And the Indian Subcontinent, Including Pakistan, Nepal, Bhutan, Bangladesh, Sri Lanka & the Maldives* (North Clarendon, VT: Tuttle Publishing, 2002).

43 *Suydam had watched a young boy approach their camp:* Cutting, *The Fire Ox and Other Years*.

43 *"Row after row of grinning skulls":* Ibid.

44 *"The reason":* Ibid.

45 *Herbert had caught two bright-orange ducks:* Stevens, *Through Deep Defiles*.

46 *At first Herbert was shocked:* Ted and Kermit Roosevelt journal entries, 1929, KBRP and TRJP.

47 *"band of eight hundred Tibetan marauders":* Theodore Roosevelt and Kermit Roosevelt, *Trailing the Giant Panda* (New York: Charles Scribner's Sons, 1929).

48 *Deforestation had run rampant:* Meng Zhang, *Timber and Forestry in Qing China: Sustaining the Market* (Seattle: University of Washington Press, 2021).

48 *"Is it coincidence":* Kermit Roosevelt journal entry, 1929, KBRP.

49 *Barely visible through the mist was Mount Satseto:* Lucien Miller, ed., *South of the Clouds: Tales from Yunnan* (Seattle: University of Washington Press, 2016).

NOTES

49 *Bitter escarpments:* Victoria Sackville-West, *The Garden* (Princeton, NJ: Princeton University Press, 1918).

50 *"Mules are gone":* T. Roosevelt and K. Roosevelt, *Trailing the Giant Panda.*

51 *"fine features and rosy cheeks":* Ibid.

51 *no one could be sure which peak . . . was truly the highest:* Walt Unsworth, *Everest: A Mountaineering History* (Boston: Houghton Mifflin, 1981).

51 *The summit would not officially be reached:* Ibid.

52 *he called him "Bob," after . . . a favorite children's book:* Alfred Ollivant, *Owd Bob: The Grey Dog of Kenmuir* (Oxford: Oxford University Press, 1898).

52 *"gloomy forebodings":* T. Roosevelt and K. Roosevelt, *Trailing the Giant Panda.*

52 *"Tibetans . . . seem impervious to the wind":* Ibid.

52 *making way for the Dalai Lama to return:* Melvyn C. Goldstein, *A History of Modern Tibet, 1913–1951: The Demise of the Lamaist State* (Berkeley: University of California Press, 1989).

52 *"I, too, returned safely to my rightful":* Tsepon W. D. Shakabpa, *Tibet: A Political History* (New Haven, CT: Yale University Press 1967).

53 *whose leaders were "tyrannical":* Goldstein, *A History of Modern Tibet.*

53 *"region of perpetual snows":* T. Roosevelt and K. Roosevelt, *Trailing the Giant Panda.*

53 *"thundering through the tree-tops":* Ibid.

54 *What else, he wondered:* Ibid.

CHAPTER 5: HOUSE OF THE PRINCE

55 *"You wake every few moments":* Theodore Roosevelt and Kermit Roosevelt, *Trailing the Giant Panda* (New York: Charles Scribner's Sons, 1929).

55 *Their bodies heaved in desperation for . . . oxygen:* Erik R. Swenson and Peter Bärtsch, eds., *High Altitude: Human Adaptation to Hypoxia* (New York: Springer, 2013).

NOTES

56 *"What might be an easy climb at 10,000 feet"*: T. Roosevelt and K. Roosevelt, *Trailing the Giant Panda*.

56 *sometimes referred to as China's Grand Canyon*: Robert Marks, *China: Its Environment and History* (Lanham, MD: Rowman & Littlefield, 2012).

57 *Locals had talked about this . . . "mythical wild goat"*: Achyut Aryal, "Status and Conservation of Himalayan Serow (*Capricornis sumatraensis* ssp. *thar*) in Annapurna Conservation Area of Nepal," BRTF Nepal: A Report Submitted to the Rufford Small Grant for Nature Conservation, UK, and the People's Trust for Endangered Species, 2008.

57 *Roy Chapman Andrews . . . had shot a specimen of the serow*: Roy Chapman Andrews and Yvette Borup Andrews, *Camps and Trails in China: A Narrative of Exploration, Adventure, and Sport in Little-Known China* (New York: D. Appleton, 1918).

57 *Even better were the pictures that Roy's wife*: Ibid.

57 *"Physically and intellectually"*: "Scientist Andrews Will Try Again to Mix Marriage and Exploration," *Laredo Times*, June 16, 1935.

57 *"Much can be done by law"*: Theodore Roosevelt, *Theodore Roosevelt: An Autobiography* (New York: Charles Scribner's Sons, 1913).

58 *Most mammals, except for humans, use smell*: Tristram D. Wyatt, *Pheromones and Animal Behaviour: Communication by Smell and Taste* (Cambridge, UK: Cambridge University Press, 2003).

58 *Few social behaviors are as important as territoriality*: Ibid.

58 *In the West, Muli had been dubbed "the Lost Kingdom"*: Joseph Rock, "Land of the Yellow Lama: National Geographic Society Explorer Visits the Strange Kingdom of Muli, Beyond the Likiang Snow Range of Yunnan, China," *National Geographic* 47 (1924).

58 *The eastern ridge of the Himalayas created a physical barrier*: Thomas Mullaney, *Coming to Terms with the Nation: Ethnic Classification in Modern China* (Berkeley: University of California Press, 2011).

59 *the mysterious paradise of Shangri-La*: James Hilton, *Lost Horizon* (New York: Macmillan, 1933).

59 *"One of the least-known spots"*: Rock, "Land of the Yellow Lama."

60 *"stiffer than any Argentine steer in cold storage"*: T. Roosevelt and K. Roosevelt, *Trailing the Giant Panda*.

NOTES

60 *a rhythm that . . . kept time for the lamas:* Francis H. Nichols, "Lamasery Life," *Bulletin of the American Geographical Society* 47, no. 2 (1915).

61 *"sinking his sharp teeth":* T. Roosevelt and K. Roosevelt, *Trailing the Giant Panda.*

61 *"a sacred sign of the buddha":* Nichols, "Lamasery Life."

61 *often serving as the only doctors:* Xiaobai Hu, "Ruling the Land of the Yellow Lama: Religion, Muli, and Geopolitics in the 17th Century Sino-Tibetan Borderland," Chinese Studies in History 52, no. 2 (2019).

61 *"I'm part lama now":* T. Roosevelt and K. Roosevelt, *Trailing the Giant Panda.*

61 *Wearing yellow was a reminder to never let go:* Nichols, "Lamasery Life."

62 *glowed eerily "ghost-like":* T. Roosevelt and K. Roosevelt, *Trailing the Giant Panda.*

63 *the image shows the six realms of existence:* Tenzin Gyatso, the Fourteenth Dalai Lama, *The Meaning of Life: Buddhist Perspectives on Cause and Effect* (Somerville, MA: Wisdom Publications, 1992).

64 *liberation from the cycle:* Ibid.

66 *"Who is President Roosevelt?":* T. Roosevelt and K. Roosevelt, *Trailing the Giant Panda.*

CHAPTER 6: SOUTH OF THE CLOUDS

67 *"It has always been remote and inaccessible":* Peter Matthiessen, *The Snow Leopard* (New York: Viking Press, 1978).

68 *accused him of being a "weakling":* Candice Millard, *The River of Doubt: Theodore Roosevelt's Darkest Journey* (New York: Doubleday, 2005).

69 *"The liquid that we tasted was so strong":* Theodore Roosevelt and Kermit Roosevelt, *Trailing the Giant Panda* (New York: Charles Scribner's Sons, 1929).

70 *"really sorry to leave":* Ibid.

71 *could also be found in Harvard's Arnold Arboretum:* Ernest H. Wilson,

NOTES

"A Phytogeographical Sketch of the Ligneous Flora of Formosa," *Journal of the Arnold Arboretum* 2, no. 1 (July 1920).

71 *This was where in 1919:* "A Giant Panda," *Natural History* 19 (October–May 1919).

71 *"Slight as our chance might be":* T. Roosevelt and K. Roosevelt, *Trailing the Giant Panda*.

72 *"a pretty little hollow"; "shrill" singing; enter a "harsh wasteland":* Ibid.

73 *"One side would be cooked":* Ibid.

73 *"We were almost exhausted":* Jack Young, "Hunting the Giant Panda with the Roosevelts in Central Asia," *China Weekly Review*, June 29, 1929.

74 *"Personally I did not blame them":* T. Roosevelt and K. Roosevelt, *Trailing the Giant Panda*.

74 *stung their faces "like birdshot":* Ibid.

75 *Members of this family of flowering plants play a critical role:* Jennifer Michel et al., "A Review on the Potential Use of Medicinal Plants from Asteraceae and Lamiaceae Plant Family in Cardiovascular Diseases," *Frontiers in Pharmacology* 5, no. 11 (2020).

75 *The vast majority of the world, more than 80 percent:* World Health Organization, "WHO Establishes the Global Centre for Traditional Medicine in India," March 25, 2022.

76 *"Ash white gnarled trunks and limbs":* T. Roosevelt and K. Roosevelt, *Trailing the Giant Panda*.

77 *"It is a truth universally acknowledged":* Jane Austen, *Pride and Prejudice* (London: T. Egerton, 1813).

77 *"The lack of power to take joy in outdoor nature":* Theodore Roosevelt, "The People of the Pacific Coast," *New Outlook* 99 (September–December 1911).

77 *"It was a ridiculous little animal":* T. Roosevelt and K. Roosevelt, *Trailing the Giant Panda*.

77 *"a fine figure of a man":* Ibid.

78 *"Practically all natives steal"; "Wisps of pale green moss":* Ibid.

79 *The young needles were packed with vitamin C:* D. J. Durzan, "Arginine, Scurvy and Cartier's 'Tree of Life,'" *Journal of Ethnobiology and Ethnomedicine* 5 (2009).

80 *Barley is one of the few crops:* Susie Neilson, "Tsampa: The Tibetan Cereal That Helped Spark an Uprising," NPR, June 23, 2019.
80 *Tsampa became a unifying force . . . of Tibetan identity:* Ibid.
81 *"Looking forward from some pass":* T. Roosevelt and K. Roosevelt, *Trailing the Giant Panda.*
82 *"It is most disconcerting"; "dirty and shabby"; "The man you bought them from":* Ibid.
83 *had recently been translated into English in 1927:* W. Y. Evans-Wentz, *The Tibetan Book of the Dead* (Oxford: Oxford University Press, 1927).
83 *"Are you oblivious to the sufferings of birth, old age, sickness, and death?":* Ibid.
83 *"Life is continually arising, dwelling, ceasing, and arising":* Pema Chödrön, *Embracing the Unknown: Life Lessons from* The Tibetan Book of the Dead (Louisville, CO: Sounds True, 2019).
84 *It was a sambar, a species of deer designated* Rusa unicolor*:* I. A. Vislobokova, "Giant Deer: Origin, Evolution, Role in the Biosphere," *Paleontological Journal* 46, no. 7 (2012).
84 *importance of the Asian continent in the evolution of giant deer:* Ibid.
84 *"They look like a cross of moose":* Dale Bowman, "The Outdoors Component of Museum Campus: The Field, the Shedd, and Lake Michigan," *Chicago Sun-Times,* January 17, 2018.
85 *"We had secured the best maps":* T. Roosevelt and K. Roosevelt, *Trailing the Giant Panda.*
85 *"There are those who claim";"Let's lengthen our marches"; "Their sputtering light":* Ibid.
87 *Sleep tends to be fragmented, lingering in the lightest sleep stages:* Stephen Bezruchka, *Altitude Illness: Prevention & Treatment* (Seattle: Mountaineers Books, 2005).
87 *even a slight reduction in the lungs' oxygen concentration:* Ibid.
88 *"worst night" of their lives:* T. Roosevelt and K. Roosevelt, *Trailing the Giant Panda.*

NOTES

CHAPTER 7: FORGE OF ARROWS

90 *"We cling to such extreme moments"*: Peter Matthiessen, *The Snow Leopard* (New York: Viking Press, 1978).

90 *"tucked deep down in the valley"*: Theodore Roosevelt and Kermit Roosevelt, *Trailing the Giant Panda* (New York: Charles Scribner's Sons, 1929).

90 *Robert Cunningham had studied at the Manchester Royal Eye Hospital:* Gerald H. Anderson, *Biographical Dictionary of Christian Missions* (Grand Rapids, MI: Wm. B. Eerdmans, 1998).

91 *Jack could expertly mimic the accents:* Michael Kiefer interview with Tai Jack Young, January 12, 1990, MKC.

91 *He was left in an uncomfortable limbo:* Ibid.

91 *"These Tibetans are men of splendid physique"*: Jack Young, "Hunting the Giant Panda with the Roosevelts in Central Asia," *China Weekly Review*, June 29, 1929.

91 *"The city, 5400 feet high"*: Suydam Cutting, *The Fire Ox and Other Years* (New York: Charles Scribner's Sons, 1940).

92 *"bowls made from human skulls"*: T. Roosevelt and K. Roosevelt, *Trailing the Giant Panda*.

93 *"a low, demoralized, sensual, avaricious class"*: Jeff Kyong-McClain, "Reaching Tibet: Anglophone Protestant Missionaries and the Chinese Civilizing Mission," *International Institute for Asia Studies Newsletter*, Autumn 2021.

93 *the lamas in this border region had risen up against the Chinese:* Wu Yuzhang, *Recollections of the Revolution of 1911: A Great Democratic Revolution of China* (Forest Grove, OR: University Press of the Pacific, 2001).

93 *"Each lama . . . on entering or leaving"*: T. Roosevelt and K. Roosevelt, *Trailing the Giant Panda*.

93 *"What do you think of the New China movements?"*: Young, "Hunting the Giant Panda."

93 *He was "sympathetic"*: Ibid.

94 *had a "fine, full-bodied" vintage:* Cutting, *The Fire Ox and Other Years*.

95 *"This definitely discredited our original information"*: T. Roosevelt and K. Roosevelt, *Trailing the Giant Panda*.

NOTES

95 *"We've clearly been led on a wild-goose chase"*: Ibid.
95 *"It's not hunting weather"*: Ibid.
96 *fresh bear tracks in the snow*: Cutting, *The Fire Ox and Other Years*.
96 *Tracks, on the other hand, are steady and dependable*: O. C. Lempfert, *Paw Prints: How to Identify Rare and Common Mammals by Their Tracks* (Ann Arbor, MI: University of Michigan Press, 1972).
96 *The shape of the paw print, its measurements . . . were all indicative*: Ibid.
96 *Notorious among these was the Sankebetsu brown bear incident*: John Knight, ed., *Natural Enemies: People-Wildlife Conflicts in Anthropological Perspective* (London: Routledge, 2000).
96 *"worst bear attack in Japanese history"*: Mark Brazil, *Japan: The Natural History of an Asian Archipelago* (Princeton, NJ: Princeton University Press, 2022).
97 *"unhabitable for winter"*: David Laichtman, "Onikuma: The Sankebetsu Brown Bear Incident and Japanese Modernity" (master's thesis, Arizona State University, May 2020).
97 *Its jaws can exert an astonishing twelve hundred psi*: Julia Taubmann et al., "Status Assessment of the Endangered Snow Leopard *Panthera uncia* and Other Large Mammals in the Kyrgyz Alay, Using Community Knowledge Corrected for Imperfect Detection," *Oryx* 50, no. 2 (2016).
97 *A solitary swipe of its massive claws*: Louis Liebenberg et al., *Practical Tracking: A Guide to Following Footprints and Finding Animals* (Mechanicsburg, PA: Stackpole Books, 2010).
98 *The mammal is a scavenger*: Dave Taylor, *Black Bears: A Natural History* (Toronto: Fitzhenry & Whiteside, 2006).
98 *In 1902, President Roosevelt was hunting black bear*: Dianne D. Glave and Mark Stoll, eds., *To Love the Wind and the Rain: African Americans and Environmental History* (Pittsburgh: University of Pittsburgh Press, 2005).
98 *calling it "unsportsmanlike"*: Ibid.
98 *"replacing dolls with toy bears"*: "Forever Young Teddy Bear Still America's Favorite Toy," *Index-Journal* (Greenwood, SC), December 14, 1982.
98 *the largest carnivore on land*: James Raffan, *Ice Walker: A Polar Bear's*

NOTES

Journey Through the Fragile Arctic (New York: Simon & Schuster, 2020).

98 *The animal is by far the most lethal of the three:* Ibid.

99 *The fur is colorless; The color is essential to life in the Arctic Circle:* Ibid.

99 *"neatly prepared" for the museum:* Young, "Hunting the Giant Panda."

100 *The animal was native to the Himalayas:* Achyut Aryal et al., "Is Trophy Hunting of Bharal (Blue Sheep) and Himalayan Tahr Contributing to Their Conservation in Nepal?" *Hystrix* 26, no. 2 (2015).

100 *two "fine rams" fell:* T. Roosevelt and K. Roosevelt, *Trailing the Giant Panda*.

101 *"They're bloodthirsty savages":* Ibid.

101 *"I only hope":* Eleanor B. Roosevelt, *Day Before Yesterday* (Garden City, NY: Doubleday, 1959).

CHAPTER 8: COMPLETE HEAVEN

102 *"The rainy season starts in April":* Theodore Roosevelt and Kermit Roosevelt, *Trailing the Giant Panda* (New York: Charles Scribner's Sons, 1929).

103 *"The land of savages":* Ibid.

103 *"We're going to meet you in Yachow":* Jack Young, "Hunting the Giant Panda with the Roosevelts in Central Asia," *China Weekly Review*, June 29, 1929.

104 *"I was in a quandary":* Herbert Stevens, *Through Deep Defiles to Tibetan Uplands: The Travels of a Naturalist from the Irrawaddy to the Yangtse* (London: H. F. G. Witherby, 1934).

104 *"The lazy indolent feeling of Spring was in the air":* T. Roosevelt and K. Roosevelt, *Trailing the Giant Panda*.

104 *"the pear, peach, and apricot trees":* Suydam Cutting, *The Fire Ox and Other Years* (New York: Charles Scribner's Sons, 1940).

105 *China is home to an astonishing 10 percent of the world's flowering plant species:* Li-Min Lu et al., "Evolutionary History of the Angiosperm Flora of China," *Nature* 554 (2018).

105 *referred to as a "third pole"*: Yao Tandong, "From Tibetan Plateau to Third Pole and Pan-Third Pole," *Bulletin of Chinese Academy of Sciences* 32, no. 9 (2017).
105 *the land was spared from the dramatic ecological turnover*: Ibid.
105 *described as akin to a "museum"*: Stevens, *Through Deep Defiles*.
105 *"seeking strange flowers"*: Joseph Rock, "Seeking Strange Flowers, in the Far Reaches of the World," *Boston Evening Transcript,* September 14, 1927.
105 *still unfolds its creamy yellow blossoms:* Peter DeMarco, "Seeking Strange Flowers," *Harvard Magazine*, September–October 2015.
106 *"They are of all ages and both sexes":* T. Roosevelt and K. Roosevelt, *Trailing the Giant Panda*.
106 *"It is nothing uncommon":* Cutting, *The Fire Ox and Other Years*.
107 *"most satisfactory method":* T. Roosevelt and K. Roosevelt, *Trailing the Giant Panda*.
108 *"deep circles under their eyes":* Cutting, *The Fire Ox and Other Years*.
108 *"It's like prohibition in the United States":* T. Roosevelt and K. Roosevelt, *Trailing the Giant Panda*.
108 *"What the physical drudgery begins":* Cutting, *The Fire Ox and Other Years*.
108 *"Have you seen this?":* Ibid.
108 *"I trapped one in a pitfall beyond Muping":* Ibid.
109 *spotted "large heaps of dung":* George B. Schaller, *The Last Panda* (Chicago: University of Chicago Press, 1994).
109 *"It seemed incongruous":* T. Roosevelt and K. Roosevelt, *Trailing the Giant Panda*.
110 *a "particularly aggravated form":* Ibid.
110 *the massive, uncomfortable growths that hung:* V. Ramalingaswami, "Endemic Goiter in Southeast Asia: New Clothes on an Old Body," *Annals of Internal Medicine* 78, no. 2 (1973).
111 *Suddenly they were surrounded by twenty bandits:* Young, "Hunting the Giant Panda."
111 *He ignored them:* Michael Kiefer interview with Tai Jack Young, January 12, 1990, MKC.
111 *Jack knew he and his guides were outnumbered:* T. Roosevelt and K. Roosevelt, *Trailing the Giant Panda*.

NOTES

112 *He heard noises, maybe even shouting:* Michael Kiefer interview with Tai Jack Young, January 12, 1990, MKC.

112 *Perched on either side of the path were dozens of men:* Ibid.

112 *She was the bait, meant to lure caravans:* Cutting, *The Fire Ox and Other Years.*

112 *they resembled a small army:* Michael Kiefer interview with Tai Jack Young, January 12, 1990, MKC.

113 *All proved bandits must have their heads cut off without trial:* T. Roosevelt and K. Roosevelt, *Trailing the Giant Panda.*

113 *local warlords, or* junfa, *seized large swaths of the countryside:* Donovan C. Chau and Thomas M. Kane, eds., *China and International Security: History, Strategy, and 21st-Century Policy* (New York: Bloomsbury, 2014).

113 *"the darkest corner in twentieth-century Chinese history":* Ibid.

113 *The fighting lasted three days:* Young, "Hunting the Giant Panda."

114 *"Muping . . . is one of those strange":* T. Roosevelt and K. Roosevelt, *Trailing the Giant Panda.*

114 *lived in "poverty and seclusion"; "astonished" at his lack of information; although "shot only rarely":* Ibid.

115 *The Chinese government had killed or evicted:* Anne-Marie Blondeau and Katia Buffetrille, eds., *Authenticating Tibet: Answers to China's 100 Questions* (Berkeley: University of California Press, 2008).

115 *accused of "barbarism":* Ibid.

115 *Buddhism had originally come to China:* Xiaoqun Xu, "The Dilemma of Accommodation: Reconciling Christianity and Chinese Culture in the 1920s," *Historian* 60, no. 1 (Fall 1997).

115 *The Christian missionaries had been allowed into China:* Ibid.

115 *"Please help us":* T. Roosevelt and K. Roosevelt, *Trailing the Giant Panda.*

116 *"I am persuaded that their fear is groundless"; "Are you sure you're a Christian":* Ibid.

117 *"What does the United States think":* Ibid.

117 *the "usual affair":* Cutting, *The Fire Ox and Other Years.*

117 *this was clearly a rare treat:* "Wild Monkeys Hunted for Food and Medicine for Centuries on Mainland," *South China Morning Post,*

NOTES

July 9, 2007; Richard C. Paddock, "Monkey Brains on the Menu," *Los Angeles Times,* February 25, 2003; Wenying Xu, *Eating Identities: Reading Food in Asian American Culture* (Honolulu: University of Hawai'i Press, 2007).

CHAPTER 9: KINGDOM OF THE GOLDEN MONKEY

118 *"Just a few hours ago":* Theodore Roosevelt and Kermit Roosevelt, *Trailing the Giant Panda* (New York: Charles Scribner's Sons, 1929).

119 *a monkey with rich golden fur and a bright-blue face:* George B. Schaller, *A Naturalist and Other Beasts: Tales from a Life in the Field* (Pittsburgh: University of Pennsylvania, 2007).

119 *"Colonel, how I envy you going where those beautiful monkeys are!":* T. Roosevelt and K. Roosevelt, *Trailing the Giant Panda.*

119 *They had first been described in 1897:* A. Milne-Edwards, "Note sur quelques mammifères du Thibet oriental," *Comptes rendus hebdomadaires des séances de l'Académie des Sciences* 70 (1870).

120 *help them withstand freezing temperatures:* Jennifer S. Holland, "The Monkey Who Went into the Cold," *National Geographic*, February 2011.

120 *regenerating the forest simply by their presence:* Hui Yao et al., "Endozoochorous Seed Dispersal by Golden Snub-Nosed Monkeys in a Temperate Forest," *Integrative Zoology* 16, no. 1 (2021).

120 *"blanket-like foliage" around them:* T. Roosevelt and K. Roosevelt, *Trailing the Giant Panda.*

120 *"It sounded like a miniature battle":* Ibid.

122 *"the horns of a wildebeest, the nose of a moose, and the body of a bison":* San Diego Zoo takin exhibit literature, San Diego, CA.

122 *Large groups of females live together with single breeding males:* Xiao-Guang Qi et al., "Satellite Telemetry and Social Modeling Offer New Insights into the Origin of Primate Multilevel Societies," *Nature Communications* 5 (2014).

122 *the male offspring are pushed out; To return . . . they must challenge one of the alpha males:* Ibid.

123 *He lectured the hunters about their shooting; They brought the little*

NOTES

     monkey back to camp: T. Roosevelt and K. Roosevelt, *Trailing the Giant Panda*.
124 *"We wanted to be full partners in the first panda"; "The little monkey . . . we tried to keep alive"; "At once we found ourselves in the densest jungle"*: Ibid.
125 *"It's time to head back to camp"*: Ibid.
126 *"coated with ice and choked with boulders"; "Every step I slipped"; "The moon was up"*: Ibid.
127 *The mature bamboo was speckled with a paintbox of fungi*: N. Jiyas et al., "Mechanical Superiority of *Pseudoxytenanthera* Bamboo for Sustainable Engineering Solutions," *Scientific Reports* 13 (2023).
127 *it is a grass that is so hardy*: Ibid.
128 *Finally, an extra digit emerged in the wrist*: H. Endo et al., "Role of the Giant Panda's 'Pseudo-Thumb,'" *Nature* 397 (1999).
128 *"Ted," Kermit suddenly yelled*: T. Roosevelt and K. Roosevelt, *Trailing the Giant Panda*.
128 *"The crackling sound of our progress"*: Suydam Cutting, *The Fire Ox and Other Years* (New York: Charles Scribner's Sons, 1940).
128 *"From dawn to dusk"*: Ibid.
129 *"The day was merely a repetition"; "If the game ever did come"*: T. Roosevelt and K. Roosevelt, *Trailing the Giant Panda*.
130 *"slept the heavy sleep of exhaustion"; "Each day was a repetition"; "the bear's habit of measuring itself"*: Ibid.
130 *a behavior typical of the brown and black bears*: A. Sergiel et al., "Histological, Chemical and Behavioural Evidence of Pedal Communication in Brown Bears," *Scientific Reports* 7 (2017).
131 *The secretions contain a wide range of compounds; They can discern age, gender, health status, and diet; A female is far more likely to mate*: Yonggang Nie et al., "Giant Panda Scent-Marking Strategies in the Wild: Role of Season, Sex and Marking Surface," *Animal Behaviour* 84, no. 1 (2012).
131 *"So much for the panda"*: T. Roosevelt and K. Roosevelt, *Trailing the Giant Panda*.

NOTES

CHAPTER 10: TEMPLE OF HELL

132 *"They were making speeches":* Theodore Roosevelt and Kermit Roosevelt, *Trailing the Giant Panda* (New York: Charles Scribner's Sons, 1929).

133 *Ancient China had become known for the "five punishments":* Xiaoqun Xu, *Heaven Has Eyes: A History of Chinese Law* (Oxford: Oxford University Press, 2020).

133 *a young woman named Chunyu Tiying wrote a letter to the emperor:* Barbara Bennett Peterson, ed., *Notable Women of China: Shang Dynasty to the Early Twentieth Century* (Abingdon, UK: Taylor & Francis, 2016).

133 *"It's a sad commentary on conditions":* T. Roosevelt and K. Roosevelt, *Trailing the Giant Panda*.

133 *the death penalty was experiencing a resurgence in the 1920s:* Maurice Chammah, *Let the Lord Sort Them: The Rise and Fall of the Death Penalty* (New York: Crown, 2021).

133 *an average of 167 executions . . . two-thirds of whom were Black citizens:* Robert Bohm, *DeathQuest: An Introduction to the Theory and Practice of Capital Punishment in the United States* (London: Anderson Publishing, 1999).

134 *its stomach contents, which contained concentrated, pure acid:* Antoni Margalida et al., "Diet and Food Preferences of the Endangered Bearded Vulture *Gypaetus barbatus:* A Basis for Their Conservation," *Ibis* 151, no. 2 (2009).

134 *The birds will pick up pieces of the skeleton:* Ibid.

135 *"When we trekked up":* T. Roosevelt and K. Roosevelt, *Trailing the Giant Panda*.

136 *"What's this place called?"; "Opium," Hsuen replied; "bitter fighting" had recently taken place:* Ibid.

137 *"It was remarkable"; "We had changed from":* Ibid.

137 *The tea gardens were more than a thousand years old:* Chuchu Liu, "The Origin and Distribution of Tea Trees and Tea in China," *Journal of Tea Science Research* 13, no. 2 (2023).

137 *Some tea leaves were considered so precious:* Po-Yi Hung, *Tea Production, Land Use Politics, and Ethnic Minorities: Struggling over*

NOTES

Dilemmas in China's Southwest Frontier (New York: Palgrave Macmillan, 2015).

138 *"The Holy Rollers"*: Paul Salopek, "Looking for Muli," *National Geographic*, May 25, 2022.

138 *"As soon as a man was properly trained"*: Suydam Cutting, *The Fire Ox and Other Years* (New York: Charles Scribner's Sons, 1940).

138 *"State Chenry Stimson"*: T. Roosevelt and K. Roosevelt, *Trailing the Giant Panda*.

138 *"Yachow proved the best . . . bazaar"*: Ibid.

139 *who became the most devoted addicts of the drug*: Ibid.

139 *unprepared for the scale of the devastation in Yachow*: Yangwen Zheng, "The Social Life of Opium in China, 1483–1999," *Modern Asian Studies* 37, no. 1 (2003).

139 *The success stories were so few*: Julia Lovell, *The Opium War: Drugs, Dreams, and the Making of China* (London: Pan Macmillan, 2011).

140 *They had many names*: Stevan Harrell, ed., *Perspectives on the Yi of Southwest China* (Berkeley: University of California Press, 2001).

140 *Their society was insular*: Alain Y. Dessaint, *Minorities of Southwest China: An Introduction to Yi (Lolo) and Related Peoples* (New Haven, CT: HRAF Press, 1980).

140 *Yi society was ruled by a strict caste system*: Ibid.

140 *women had few rights and did not serve as hunters or guides*: Ibid.

141 *the "land of the Lolos"*: T. Roosevelt and K. Roosevelt, *Trailing the Giant Panda*.

141 *"they sang so pleasantly"*: T. Roosevelt and K. Roosevelt, *Trailing the Giant Panda*.

143 *"The cold and sleet had proven too much"*: Ibid.

143 *The party had collected five thousand bird skins*: "Annual Report of the Director," *Field Museum of Natural History–Reports*, vol. 8, 1929–1930.

143 *"It appeared to be a hopeless quest"*: Cutting, *The Fire Ox and Other Years*.

143 *"I'll be glad to get rid of this"*: T. Roosevelt and K. Roosevelt, *Trailing the Giant Panda*.

144 *the flowers "were a mass of color"*: Ibid.

NOTES

144 *Mulberry leaves are the sole diet for silkworms:* Ben Marsh, *Unravelled Dreams: Silk and the Atlantic World, 1500–1840* (Cambridge, UK: Cambridge University Press, 2020).

144 *the empress Leizu was sitting under a mulberry tree:* Ibid.

144 *"queen of fabrics":* W. D. Darby, "Silk, the Queen of Fabrics, Has Played a Moving Part in History," *Dry Goods Economist*, January 14, 1922.

145 Bombyx mori *can survive only in captivity:* A. H. Jingade et al., "A Review of the Implications of Heterozygosity and Inbreeding on Germplasm Biodiversity and Its Conservation in the Silkworm, *Bombyx mori*," *Journal of Insect Science* 11, no. 1 (2011).

145 *The adults have lost the ability to fly:* Ibid.

145 *"My time was short":* Herbert Stevens, *Through Deep Defiles to Tibetan Uplands: The Travels of a Naturalist from the Irrawaddy to the Yangtse* (London: H. F. G. Witherby, 1934).

146 *"Is it possible to see through people?":* Ibid.

147 *"X-rays!":* Ibid.

147 *"What's going on out there?":* T. Roosevelt and K. Roosevelt, *Trailing the Giant Panda*.

147 *Smallpox was a devastating and highly contagious virus:* Jennifer Lee Carrell, *The Speckled Monster: A Historical Tale of Battling Smallpox* (New York: Penguin, 2003).

147 *In 1796 in England, physician and scientist Edward Jenner:* Ibid.

147 *China had been inoculating against smallpox as early as 200 BC:* Michael Bennett, *War Against Smallpox: Edward Jenner and the Global Spread of Vaccination* (Cambridge, UK: Cambridge University Press, 2020).

148 *"There are times":* T. Roosevelt and K. Roosevelt, *Trailing the Giant Panda*.

148 *"Last night," he explained:* Ibid.

149 *The bear . . . "was a* beishung*":* Ibid.

NOTES

CHAPTER 11: LAND OF THE YI

150 *It began when Theodore Roosevelt hid in the family's vast Oyster Bay home*: Hermann Hagedorn, *The Roosevelt Family of Sagamore Hill* (New York: Macmillan, 1954).

150 *He, in turn, would roar as loudly as possible*: Diary entries in TRC, TRJP, and KBRP.

150 *he loved to play with his sons whenever he was home*: Ibid.

150 *"Your loving father"*: Multiple letters from Theodore Roosevelt to Ted Roosevelt Jr. and Kermit Roosevelt, TRC.

152 *"Why?" Kermit asked; a "supernatural being"; "We determined to approach"*: Theodore Roosevelt and Kermit Roosevelt, *Trailing the Giant Panda* (New York: Charles Scribner's Sons).

154 *"It's your beards"; "It was here a month ago*: Ibid.

155 *"Did you ever kill a beishung?"; "Yes, ten hunters"*: Ibid.

156 *"I never had such dreams"*: Letter from Kermit Roosevelt to Belle Roosevelt, 1929, KBRP.

157 *"Drenched by rain and soaked by snow"; "We did not want to shoot pig"*: T. Roosevelt and K. Roosevelt, *Trailing the Giant Panda*.

158 *"Hsuen's choice of words"; "The headman's got a ghost"; "We were a bedraggled lot"*: Ibid.

159 *were only seen "occasionally"; "At length the valley opened up"; "dense fog soon shrouded the valley"*: Ibid.

160 *"It's impossible to hunt in this weather"; "conventional Christmas garb"; "a cheerless morning"*: Ibid.

161 *Growing for centuries, they "walk" through the jungle*: Ben Crair, "The Biggest Tree Canopy on the Planet Stretches Across Nearly Five Acres," *Smithsonian*, April 2017.

162 *"They are panda tracks"*: T. Roosevelt and K. Roosevelt, *Trailing the Giant Panda*.

162 *"traveling along in a leisurely fashion"; "The bamboo jungle"*: Ibid.

163 *a "nest of bamboo"*: Ibid.

164 *the "old lion"*: Patricia O'Toole, *When Trumpets Call: Theodore Roosevelt After the White House* (New York: Simon & Schuster, 2005).

164 *"I seek my father in the wild places"*: Kermit Roosevelt diary entry, 1924, KBRP.

NOTES

164 *A strange "clicking chirp"*: T. Roosevelt and K. Roosevelt, *Trailing the Giant Panda*.

164 *"black spectacles" around the eyes*: Ibid.

165 *"He seemed very large"; "vanish like smoke in the jungle"; "We fired simultaneously"; "Although the Lolos had all along told us"; "We knew he was ours"*: Ibid.

CHAPTER 12: THE HALL OF ASIAN MAMMALS

167 *"We had hunted hard and long"*: Theodore Roosevelt and Kermit Roosevelt, *Trailing the Giant Panda* (New York: Charles Scribner's Sons, 1929).

168 *"A deeply interested group"; "no trace of the panda having varied"; "He's a gentleman"*: Ibid.

169 *"It's not a savage animal"; none . . . "would touch a morsel"; "all-embracing ceremony of purification"*: Ibid.

170 *"a gold ring set with turquoise"; "black and white beishungs"; she was "really embarrassed"; "When we bade her goodbye"; "It seemed almost unbelievable"*: Ibid.

172 *Hendee had traveled to Indochina*: Alfred M. Bailey, "Russell W. Hendee," *Condor* 32, no. 3 (1930).

173 *He had trained at the University of Iowa*: Ibid.

174 *"most perilous time of my life"*: Ted Roosevelt to Seymour Halpern, 1929, TRJP.

174 *"Together we had shivered"*: T. Roosevelt and K. Roosevelt, *Trailing the Giant Panda*.

174 *"thousand miles of downstream travel"*: Bailey, "Russell W. Hendee."

175 *Hendee had hired a river guide*: Ibid.

176 Plasmodium falciparum, *the parasite that causes cerebral malaria*: Randall M. Packard, *The Making of a Tropical Disease: A Short History of Malaria* (Baltimore: Johns Hopkins University Press, 2021).

176 *the parasite had developed resistance to the widely used medication*: Christopher V. Plowe, "The Evolution of Drug-Resistant Malaria" *Transactions of the Royal Society of Tropical Medicine and Hygiene* 103, no. 1 (2009).

NOTES

176 *With a leap, the young man plunged to his death:* "Brooklyn Scientist Leaps to His Death," *New York Times,* June 8, 1929.

176 *On the sidewalk below, hospital workers saw Hendee:* "Ill-Luck Delays Roosevelt Return," *New York Times,* June 23, 1929.

177 *In a small graveyard in Laos, a stone marked Hendee's final resting spot:* Author interview with member of Hendee's family, 2022.

177 *the eldest Roosevelt brother began perspiring so heavily:* Letters from Theodore Roosevelt Jr. to Eleanor B. Roosevelt, 1929, TRJP.

178 *His fever climbed to 105 degrees Fahrenheit:* Ibid.

178 *the disease had proved especially deadly after the country began to increase its rice production:* Cuong Do Manh et al., "Vectors and Malaria Transmission in Deforested, Rural Communities in North-Central Vietnam," *Malaria Journal* 9 (2010).

178 *"Never in all my life":* Eleanor B. Roosevelt, *Day Before Yesterday* (Garden City, NY: Doubleday, 1959).

179 *"rare beast of the Tibetan border":* "Panda Killed by Roosevelts Is Sacred to Asia Natives," *New York Times,* June 2, 1929.

179 *"Wild animals . . . only continue to exist":* Theodore Roosevelt, *Outdoor Pastimes of an American Hunter* (New York: Charles Scribner's Sons, 1905).

179 *"I'm not myself":* Letter from Kermit Roosevelt to Belle Roosevelt, 1929, KBRP.

179 *"the most challenging animal trophy on Earth":* Desmond and Ramona Morris, *Men and Pandas* (New York: McGraw-Hill, 1966).

180 *"panda-mania" began sweeping:* Chris Heller, "How America Fell in Love with the Giant Panda," *Smithsonian,* September 21, 2015.

180 *The Roosevelts had hastily published a book:* T. Roosevelt and K. Roosevelt, *Trailing the Giant Panda.*

181 *Ruth decided to take up her husband's cause:* Ruth Harkness, *The Lady and the Panda: An Adventure* (New York: Carrick & Evans, 1938).

181 *Harkness managed to smuggle a six-week-old cub:* Vicki Croke, *The Lady and the Panda: The True Adventures of the First American Explorer to Bring Back China's Most Exotic Animal* (New York: Random House, 2006).

181 *after paying China an "export tax" of $45:* Foster Stockwell, *West-*

NOTES

erners in China: A History of Exploration and Trade, Ancient Times Through the Present (Jefferson, NC: McFarland, 2015).

181 *The Brookfield Zoo in Chicago was not as picky*: Michael Kiefer, *Chasing the Panda: How an Unlikely Pair of Adventurers Won the Race to Capture the Mythical "White Bear"* (New York: Four Walls Eight Windows, 2002).

182 *"If it was up to me"*: Letter, Theodore Roosevelt Jr. to Archie Roosevelt, 1936, LOC.

182 *"He's my fifth cousin"*: Tim Brady, *His Father's Son: The Life of General Ted Roosevelt Jr.* (New York: Penguin, 2017).

182 *"I am tremendously relieved"*: Ibid.

183 *"I'd just as soon think of mounting"*: D. and R. Morris, *Men and Pandas*.

183 *"We were responsible"*: Kermit Roosevelt diary entry, 1936, KBRP.

183 *For a flat fee of $150*: Kiefer, *Chasing the Panda*.

185 *"Popular Panda"*: Found on the original placard for Su-Lin at the Field Museum, Chicago, 1938; FM.

CHAPTER 13: THE SUMMER WHITE HOUSE

186 *In 1904, a group of British soldiers*: Peter Hopkirk, *Trespassers on the Roof of the World: The Race for Lhasa* (London: John Murray, 1982).

187 *their expedition became one of the few to be issued a permit*: Suydam Cutting, *The Fire Ox and Other Years* (New York: Charles Scribner's Sons, 1940).

187 *"I sought no hide-and-seek"*: Ibid.

188 *the "first white Christian"*: "C. Suydam Cutting, Who Made Historic Visit to Tibet, Is Dead," *New York Times,* August 25, 1972.

188 *"Freedom was part of their tradition"*: Suydam Cutting, *The Fire Ox and Other Years*.

188 *"It is a corollary of reincarnation that all life"*: Ibid.

188 *"We shall fight on the beaches"*: Andrew Roberts, *Churchill: Walking with Destiny* (New York: Penguin, 2018).

189 Suydam started the American Committee for the Defense of Brit-

NOTES

- *ish Homes:* Malcolm Atkin, *To the Last Man: The Home Guard in War and Popular Culture* (Barnsley, UK: Pen & Sword Books, 2019).
- 189 *He loved his Tibetan dogs until the end, breeding Lhasa apsos:* American Kennel Club, *The New Complete Dog Book, 23rd Edition: Official Breed Standards and Profiles for Over 200 Breeds* (Mount Joy, PA: Fox Chapel, 2023).
- 189 *He died in 1972 at age eighty-three:* "C. Suydam Cutting," *New York Times.*
- 189 *In 1930, the Royal Geographical Society published:* Herbert Stevens, "Sketches of the Tatsienlu Peaks," *Geographical Journal* 75, no. 4 (April 1930).
- 190 *in 1934 published his own account of the trail:* Herbert Stevens, *Through Deep Defiles to Tibetan Uplands: The Travels of a Naturalist from the Irrawaddy to the Yangtse* (London: H. F. G. Witherby, 1934).
- 190 *"Long live the land of the lamas!":* Ibid.
- 190 *"Stevens has done a great deal of excellent work":* Ronald E. F. Peal, "The Chairman's Address," *Bulletin of the British Ornithologists' Club*, 111, no. 4 (1991).
- 191 *Jack came back to the United States . . . roughly $3,000 richer:* Michael Kiefer, *Chasing the Panda: How an Unlikely Pair of Adventurers Won the Race to Capture the Mythical "White Bear"* (New York: Four Walls Eight Windows, 2002).
- 191 *He joined the Explorers Club in New York; He applied to the Field Museum for a low-level position; Together, the group decided to climb Minya Konka:* Ibid.
- 192 *In return for allowing Jack and his companions to make the trip; Even this price, however, was difficult to obtain; "a headless and footless specimen, difficult to mount":* Ibid.
- 193 *Suydam, Ted, and Kermit invited the young naturalist to Oyster Bay; Su-Lin was born and raised in New York City; the American Museum of Natural History offered him $2,700:* Ibid.
- 194 *his employers at the museum berated Su-Lin for Jack's sudden absence:* Ibid.

NOTES

194 *"I fought for my country"*: Michael Kiefer interview with Tai Jack Young, January 12, 1990, MKC.

194 *He died in 2000 at ninety-one years old*: Author interview with Michael Kiefer, 2023.

194 *Jolly Young, would remember her parents fondly as explorers and scientists*: Author interview with Jolly Young, 2024.

194 *In 1932 he edited a book of war poetry*: Theodore Roosevelt, Jr., and Grantland Rice, eds., *Taps: Selected Poems of the Great War* (New York: Doubleday, Doran, 1932).

194 *It "has been in the back of my mind"*: "Colonel Roosevelt Joins Book Firm," *New York Times*, September 16, 1935.

194 *"I would feel as desolate"*: Theodore Roosevelt and Kermit Roosevelt, *East of the Sun and West of the Moon* (New York: Charles Scribner's Sons, 1926).

195 *the "only book in English"*: Karen J. Leong, *The China Mystique: Pearl S. Buck, Anna May Wong, Mayling Soong, and the Transformation of American Orientalism* (Berkeley: University of California Press, 2005).

195 *Ted continued to write his own books as well*: Theodore Roosevelt Jr., *Colonial Policies of the United States* (New York: Doubleday, Doran, 1937).

195 *"Ted's truly remarkable career"*: Eleanor B. Roosevelt, *Day Before Yesterday* (Garden City, NY: Doubleday, 1959).

195 *"overflowed into every room in the house"*: Theodore Roosevelt Jr., *All in the Family* (New York: G. P. Putnam's Sons, 1929).

196 *Kermit took an executive position with the New York Zoological Society*: William Lemanski, *Lost in the Shadow of Fame: The Neglected Story of Kermit Roosevelt, a Gallant and Tragic American* (Mechanicsburg, PA: Sudbury Press, 2012).

196 *Kermit and Suydam reunited aboard Vincent Astor's massive 263-foot yacht*: Kermit Roosevelt, "The Mountain Party on Indefatigable Island," *New York Zoological Society Bulletin* 23, no. 4 (July–August 1930).

196 *the yacht was also a "floating laboratory"*: "The Ship upon Which the President-Elect Cruises Has Made Long Voyages in the Interest of Science," *New York Times*, February 12, 1933.

NOTES

196 *The islands had been named for the massive numbers of tortoises:* Elizabeth Hennessy, *On the Backs of Tortoises: Darwin, the Galápagos, and the Fate of an Evolutionary Eden* (New Haven, CT: Yale University Press, 2019).

197 *The largest of the tortoises spanned five feet:* Ibid.

197 *"This animal . . . formerly of great food value":* K. Roosevelt, "The Mountain Party."

197 *Kermit and the zoological society had determined that a breeding program:* Ibid.

197 *"the Galapágos Islands where they helped":* "C. Suydam Cutting," *New York Times.*

197 *Kermit became president of the Audubon Society:* Lemanski, *Lost in the Shadow of Fame.*

198 *he boldly called for a national closure of duck hunting:* Theodore W. Cart, "'New Deal' for Wildlife: A Perspective on Federal Conservation Policy, 1933–40," *Pacific Northwest Quarterly* 63, no. 3 (July 1972).

198 *Kermit began reaching out to leaders of foreign nations:* Audubon Society Archives, 1934–1944, NYPL.

198 *"All hunting must cease":* Ibid.

199 *pandas that were "dying in captivity":* Ibid.

199 *"The only solution":* Ibid.

199 *described as an "arduous journey":* Michael Kiefer, *Chasing the Panda: How an Unlikely Pair of Adventurers Won the Race to Capture the Mythical "White Bear"* (New York: Four Walls Eight Windows, 2002).

199 *The marine mammal . . . was a victim of the fur trade:* Todd McLeish, *Return of the Sea Otter: The Story of the Animal That Evaded Extinction on the Pacific Coast* (Seattle: Sasquatch Books, 2018).

199 *Then, out of nowhere, a raft of sea otters:* "The 'Extinct' Sea Otter Swims Back to Life," *Life,* June 20, 1938.

199 *In 1938, China banned all giant panda hunting:* Robert A. Montgomery, "Characteristics that make trophy hunting of giant pandas inconceivable," *Conservation Biology* Vol 34(4) 2020.

200 *The government also cracked down on exporting live animals:* Ibid.

200 *"their playful antics will bring as much joy to American children . . .":*

NOTES

"Bronx to Get New Panda: Mme. Chiang Kai-shek Allows Gift in Thanks for China Aid," *New York Times*, September 12, 1941.

200 *like the stolen animals that preceded them, the cubs wouldn't live long:* Henry Nicholls, *The Way of the Panda: The Curious History of China's Political Animal* (New York: Pegasus Books, 2011).

200 *It would stay that way until 1972:* Ibid.

EPILOGUE: TRAIL'S END

201 *a sedative used at the time to treat alcohol withdrawal:* R. N. Chopra et al., "Chloral Hydrate and Paraldehyde as Drugs of Addiction," *Indian Medical Gazette* 67, no. 9 (1932).

202 *By 1940, he had been made a colonel in the British Army:* "Kermit Roosevelt Greets His Brigade; Sees Force He Will Command in Finland," *New York Times*, May 6, 1940.

202 *noted for "helping get men and equipment":* Edward J. Renehan Jr., *The Lion's Pride: Theodore Roosevelt and His Family in Peace and War* (Oxford: Oxford University Press, 1998).

202 *"Major Kermit Roosevelt has been to me":* Martin Gilbert, ed., *The Churchill Documents, vol. 16, The Ever-Widening War 1941* (Hillsdale, MI: Hillsdale College Press, 2011).

202 *When the United States entered . . . Kermit was quick to join up:* William Lemanski, *Lost in the Shadow of Fame: The Neglected Story of Kermit Roosevelt, a Gallant and Tragic American* (Mechanicsburg, PA: Sudbury Press, 2012).

203 *"There is a universal saying to the effect":* Kermit Roosevelt, *The Long Trail* (New York: Charles Scribner's Sons, 1921).

203 *On the evening of June 3:* "Maj. Kermit Roosevelt Dies in Alaska; Son of Late President," *Washington Post*, June 6, 1943.

203 *"Sleep," his buddy replied:* Renehan, *The Lion's Pride*.

203 *"He really died five years ago":* Tim Brady, *His Father's Son: The Life of General Ted Roosevelt Jr.* (New York: Penguin, 2017).

203 *he was placed on active duty at his earnest request:* Ibid.

203 *Although he was fifty-six and walking with a cane:* Ibid.

NOTES

203 *"felt sure he would be killed"*: Joseph Balkoski, *Utah Beach: The Amphibious Landing and Airborne Operations on D-Day, June 6, 1944* (Mechanicsburg, PA: Stackpole, 2005).

203 *"We'll start the war from right here!"* Stephen E. Ambrose, *D-Day: June 6, 1944: The Climactic Battle of World War II* (New York: Simon & Schuster, 1994).

204 *The Chinese paddlefish, which could once grow to more than twenty-three feet:* Douglas Main, "The Chinese Paddlefish, One of World's Largest Fish, Has Gone Extinct," *National Geographic*, January 8, 2020.

204 *a species now critically endangered in China:* Sofia Quaglia, "Is the Yangtze River Dolphin Gone Forever?" *Discover*, June 3, 2023.

204 *The sarus crane that Herbert loved with a passion . . . is endangered:* James Harris and G. W. Archibald, "Status of and Threats to the Cranes of the World," *Proceedings of the VII European Crane Conference: Breeding, Resting, Migration and Biology,* Stralsund, Germany, 2010.

204 *The colorful hornbills . . . have also mostly disappeared:* C. E. R. Hatten et al., "Three Birds with One Stone? Sex Ratios of Seized Critically Endangered Helmeted Hornbill Casques Reveal Illegal Hunting of Males, Females and Juveniles," *Animal Conservation* 26, no. 4 (2023).

205 *Trail camera footage in Vietnam in 2014 might have spotted a glimpse:* Minh Le et al., "Discovery of the Roosevelt's Barking Deer (*Muntiacus rooseveltorum*) in Vietnam," *Conservation Genetics* 15 (2014).

205 *"It's often easier to find animals from 10 different species"*: Li You, "China's Endangered Birds Gain an Unexpected Ally," *Sixth Tone*, February 21, 2020.

205 *A 2021 expedition along the Mekong River . . . discovered 205 new species:* Mia Signs et al., "New Species Discoveries in the Greater Mekong, 2021 and 2022," *WWF–Greater Mekong*, 2023.

205 *"I can only view with irony"*: George B. Schaller, *The Last Panda* (Chicago: University of Chicago Press, 1994).

205 *in 2021, the giant panda was moved from endangered to vulnerable:* Kyle

NOTES

Obermann, "China Declares Pandas No Longer Endangered—But Threats Persist," *National Geographic*, September 1, 2021.

206 *"It is the difficult fate of this generation"*: Schaller, *The Last Panda*.

206 *its population has remained largely intact*: Tianpei Guan, "Nature Reserve Requirements for Landscape-Dependent Ungulates: The Case of Endangered Takin (*Budorcas taxicolor*) in Southwestern China," *Biological Conservation* 182 (2015).

206 *Due to the combined effects of poachers, deforestation, and disease*: S. Li et al., "Retreat of Large Carnivores Across the Giant Panda Distribution Range," *Nature Ecology & Evolution* 4 (2020).

206 *"a survey based on direct sightings"*: Ronald R. Swaisgood, "Panda Downlisted but Not Out of the Woods," *Conservation Letters* 11, no. 1 (2018).

207 *new protections have been given to elephants in Asia and Africa*: Robin Naidoo et al., "Estimating Economic Losses to Tourism in Africa from the Illegal Killing of Elephants," *Nature Communications*, November 1, 2016.

207 *Efforts are also underway to protect the world's tigers*: Charukesi Ramadurai, "As India's Project Tiger Turns 50, Hope for the Big Cat," *Christian Science Monitor*, April 4, 2023.

INDEX

Page numbers of photographs appear in italics.

Key to abbreviations and first names: AMNH = American Museum of Natural History; Herbert = Herbert Stevens; Jack = Tai Jack Young; Kermit = Kermit Roosevelt; Suydam = [Charles] Suydam Cutting; Ted = Theodore Roosevelt, Jr.; TR = Theodore Roosevelt

A
American Geographical Society, 192
American Museum of Natural History (AMNH), xv–xvii
  Andrews's serow specimen at, 57
  funds Jack's 1937 expedition, 193
  panda specimens at, 72, 192–93
Amundsen, Roald, xviii
Andrews, Roy Chapman, 57, 101
Andrews, Yvette Borup, 57
Arnold Arboretum, Harvard University, Boston, 71, 105
Astor, Vincent, 196

Audubon Society
  Bird-Lore publication, 198
  Kermit as president, 197–98
Austen, Jane, 77, 194, 203
  Pride and Prejudice, 77, 139

B
Baker, Steven, 190
bamboo, 127
  barbed spines (glochids), 162–63
  panda diet and, 127–28
  panda habitat and, 159, 165, 168
  panda scent marking, 130–31
banyan trees, *Ficus macrocarpa*, 161

INDEX

bearded vulture, 134–35
Beijing lilac, *Syringa pekinensis*, 105
bighorn sheep, *Ovis poli*, xviii–xix
black bear, *Ursus americanus*, 98
blue sheep, *Pseudois nayaur*, 100
Bronx Zoo, New York, 200
Brookfield Zoo, Chicago, 181, 183
brown bear, *Ursus arctos*, 96–98
  notorious Sankebetsu brown bear incident, 96–97
Burton, Sir Richard F.
  Personal narrative of a pilgrimage to el Medinah and Meccah, 84

C

Chang Ta Li, 65–66, 67, 71, 72
  family of, 65–66
  gifts for the Roosevelts, 70
Cherrie, George, 5
Chiang, Madame (Soong Mei-Ling), 194–95, 200
Chiang Kai-shek, 53
China
  "the back door" from Burma into, 2–3
  biodiversity of, 105, 205
  Buddhism in, 115
  Christian missionaries in, 115, 137–38
  civil war in, 28–29
  colonization of Tibet and, 52–53
  conflict between Christian missionaries and lamas, 93
  deforestation and destructive logging practices, 47–48
  ethnic minority groups of the Southwest, 140
  executions used by, 133
  "five punishments" of ancient China, 133
  flowering plant species of, 104–5
  foot-binding or "Lotus feet," 36–37, 48
  goiters common in, 110
  Hong Kong ceded to the British, 15
  Kermit's correspondence, advocating protection of pandas, 198, 199–200
  landmass of, xiii
  New China movements, 93
  opium in, 14–16, 108, 139
  Opium Wars of 1839, 15
  pandas absent from art of, 198–99
  permit for Jack to return in 1932, with proviso, 192
  polyandry in, 24
  Silk Road and trading, 144–45
  smallpox inoculation in, 147–48
  tea and tea porters, 23
  warlords of the countryside, 113–14

INDEX

women porters, 22–24
See also Yi people; *specific locations*
Chinese paddlefish, 204
Churchill, Winston, 202
  "fight on the beaches" speech, 188–89
Collier, Holt, 98
Colonial Policies of the United States (Ted Roosevelt), 195
Complete Heaven (Tienchuan), 102–11
  access to, as limited, 109
  description of the town, 109–10
  disease in, 110
  inhabitants, 109
  Roosevelt brothers and Suydam en route to, 103, 104
  Roosevelt brothers and Suydam rush to leave, 110
  Roosevelt brothers obtain a panda skin, 108–11, 180
  tea porters and porters toting men, *106*, 106, *107*, 107
  trail followed, 106–8
conservation, xvii
  endangered and extinct species, 204–5, 206
  habitat destruction and, 207
  hunting versus, 183, 206–7
  Kermit's commitment to, 196, 198, 199
  panda as symbol of, 207
  protection for elephants, 207
  protection for tigers, 207
  saving the Galápagos tortoise, Kermit with Suydam, 189
  saving the panda, Kermit's advocacy, 198, 199
  saving the sea otters, 199
Coolidge, Harold, Jr., 172, 173
Crim, legend of, 37–38
creature in the Mekong, 39–40
Cunningham, Robert and wife, 90–94
Cutting, [Charles] Suydam
  advocacy for endangered species, 189
  American Committee for the Defense of British Homes founded by, 189
  appearance, 9, *9*
  background, 10
  Central Asia expedition (1925), 5, 10–11
  character and personality, 9
  correspondence and exchange of dogs with the Dalai Lama, 188
  Explorers Club and, 189
  first white Christian to enter Lhasa, 188
  Kermit and, 10, 22
  Kermit's reunion with aboard the Astor yacht, 196–97
  post-expedition life, rejects hunting and studies plant life, 187
  return journeys to Tibet, 186, 187–88

Cutting, [Charles] Suydam (*cont.*)
  saving the Galápagos tortoise, with Kermit, 189, 196–97
  studying the Naga people of India, with Hutton, 43–45
  Tibetan dogs, Lhasa apsos, and, 188, 189
  wife, Helen McMahon, 188
  World War II and, 188–89
Cutting, [Charles] Suydam:
  Roosevelt brothers' panda expedition, 4, 8–9, 14, 100–101, 143
  bear tracks found by, 96, 99
  birds captured by, 42–43, 45
  expedition divides, Roosevelts go alone to Lolo land, 143
  explores the area south of Muping, 118, 128
  in Forge of Arrows (Tatsienlu), 91–92, 94
  in the Himalayas, 55, 72, 75–76, 85–86, 91–92
  hunts the panda, 95, 134
  journey to Complete Heaven, 103–9
  in Muli, 68–69, 72
  opinion of Herbert, 39
  photographs by, *44, 63, 64, 69, 71, 73, 80, 81, 86, 90, 106, 107, 134, 151, 153*
  shoots a bearded vulture, 134
  Ted and, train to Saigon and bout with malaria and dysentery, 174, 177–78
  Valley of Death and, 22, 31
  Yachow and, 134–35, 138
Cutting, Helen McMahon, 188
  first white woman to meet the Dalai Lama, 188
Cuvier, Georges-Frédéric, 35

D

Dalai Lama, 52, 187, 188
Davenat, Pere, 93
David, Armand, xiii
David, Pere, 141
Dickens, Charles
  Great Expectations, 84
dolphins, 39–41
  Irrawaddy dolphin, *Orcaella brevirostris*, 41, 204
Doubleday, Doran publishers
  publishes Ted's *Colonial Policies of the United States*, 195
  Ted becomes editor at, 194
  Ted signs Madame Chiang, 194–95

E

Everest, Mount, 51
  attempt to summit (1924), 51
  first documented ascent (1953), 51
explorers, exploration, xi
  Amundsen at the North Pole, xviii
  Andrews expedition of 1916, 57

INDEX

animals for museum collections, xii
collection of medicinal plants by, 76
disturbing delicate natural balance, 27
Fawcett in the Amazon, xviii
maps and unexplored regions, xi
museum funding for, xvii–xviii
1920s decade, xii, xvii
vanishing of, 204
See also Rock, Joseph; *specific explorers*
Explorers Club, New York, 189, 191

F

Fawcett, Percy, xviii
Field Museum of Natural History, Chicago
bighorn sheep and, xviii–xix
funding for Coolidge expedition to join the Roosevelts, 172
funding for Roosevelt brothers' 1925 expedition, xviii–xix
Kermit speaks at about the panda expedition, 179, 180
list of mammals wanted by, 56
panda exhibits, 180, *181*, 183, *184*, 185
Roosevelt brothers collect a sambar pelt for, 84–85
Roosevelt brothers' giant panda expedition and, xvii, 4, 56, 101, 107
Forge of Arrows (Tatsienlu), 71–72, 85, 89–101, 102
Christian mission run by the Cunninghams, 90–94
description of the town, 91–92
expedition members take a bath, 92
flowering plant species of, 104
food in, 94
goddess Chammo Lam Lha, 94, 102
Herbert's planned rendezvous at, 47, 103
panda skin bought by Milner in 1919 at, 72
Roosevelt brothers' failed hunt for the panda at, 94–96
Ted shoots two blue sheep rams, 100
Ted shops for treasures, 92

G

Galápagos Islands, 196
Astor's yacht *Nourmahal* and, 196–97
saving the Galápagos tortoise, 189, 196–97
giant panda, *Ailuropoda melanoleuca* (panda, panda bear), xiii–xv, *xiv*
absent from Chinese art, 198–99

INDEX

giant panda, *Ailuropoda melanoleuca* (panda, panda bear) *(cont.)*
appearance, 164–65
bamboo diet, 127–28
botanist spots dung in Min River valley (1908), 109
called *beishung* or white bear, 148, 152, 165
in Chinese literature, xiii
communication through bamboo (scent marking), 130–31
conservation and protection for, 205–6
cub purchased by Brookfield Zoo, 181, 183
cubs snatched and sold to zoos, 180, 185, 198
death in captivity, 183, 200
as endangered, 199–200
evolution of, 127–28
Field Museum's exhibits, 180, *181*, 183, *184*, 185
first panda skin seen in the West, obtained by Pere David, 141
footprint of, 161
habitat, 159, 165, 168, 200
hunting of, as ultimate challenge, 99
Jack kills a panda (1932), 192
Kermit advocates protection for, 198, 199–200
male and female genitalia hidden in cubs, 183
mystery of its habitat and behavior, xii
as mythic, xiii, xv, 36
naming of, xiii–xiv
as nonaggressive and docile, 165, 169
pair of, given by Madame Chiang to the Bronx Zoo, 200
"panda-mania" in the US, 180
panda preserves (Chengdu Research Base and Wolong Panda Center), 206
question of hibernation, 109
rarity of, xii, 152, 159, 198–99, 205, 206
Roosevelt brothers' failed hunt in the Forge of Arrows, 94–95
Roosevelt brothers' failed hunt in the Kingdom of the Golden Monkey, 124–21
Roosevelt brothers' final hunt in Lolo land and killing of, 152, 156–64
Roosevelt brothers obtain an old skin, 108–11
Roosevelt brothers' unexpected reaction to killing the panda, 167–69
scientists speculate about behavior and habitat, 99, 109
skin bought by Milner at Tatsienlu (1919), 72, 100
sighted in Lolo land, 148–49, 154–55

INDEX

as solitary, 130
unintended consequences
of the Roosevelt brothers'
expedition, 179–80, 183
Yi people consider the bear a
demi-god and do not hunt,
148–49, 155, 159
zoos' desire for, 180
Gilded Age, wealth of prominent
American families, xvii–xviii
golden snub-nosed monkey,
Rhinopithecus roxellana,
118–24
death of a newborn, impact on
the Roosevelt brothers, 124,
129
as endangered, 206
habitat, 119–20
highly desired by museums, 119
Kermit's rage at the killing of a
mother with a newborn, 123
Milne-Edwards describes, 119
Roosevelt brothers kill an
entire family, 120–21
skins collected, 123–24
Great Expectations (Dickens), 84

H
Happy Valley, 1–18
ancient trail through, 1–2
encounter with Antigone
cranes, 17–18
expedition's sojourn in Kanai
with Saw Bwa Fang Tao,
7–9, 12–17
path from Burma to China
called "the back door," 2
search for Herbert in, 4–7
Harkness, Ruth, 180–81, 182,
199
Harvard University
Herbert's Papua New Guinea
expedition and, 190
Hendee, Russell, 172–77, 190
burial stone in Laos, 177
fatal illness, 175–76
Hilton, James, *Lost Horizon*, 59
Himalayan snowball plant,
Saussurea laniceps, 75–76
Himalayas, 1, 50, 72
about, 51
cultures and minorities of, 59
detailed maps by Herbert, 189
diversity of wildlife and
plants, 59, 71, 76
highest peak unknown in
1929, 51
hot springs in, 94
Jack's wife as one of the first
women explorers, 193–94
medicinal plants found in, 76
Minya Konka mountain, size
of, 85, 191–92
Mount Everest, 51
new scientific expeditions in,
207
oldest stands of trees, 62
"region of perpetual snows,"
53
Tiger Leaping Gorge or
China's Grand Canyon, 56

INDEX

Himalayas *(cont.)*
  trails along the ridgelines, 78
  Shangri-La inspired by, 59
  spruce forest of, 78–79
Himalayas: Roosevelt brothers'
    panda expedition
  expedition ascends into, 38, 42,
    47–58
  expedition crosses the Yalong
    River, 79–81, *80*
  expedition hikes at night, en
    route to Forge of Arrows,
    85–88
  expedition maps unhelpful, 85,
    102
  expedition members' altitude
    sickness, 55, 62, 74, 87–88
  expedition near to Yunnan
    Province, 85
  expedition's mules and supplies
    lost and found, 50–52, 54, 55,
    74–75, 77–78
  expedition's route from Muli,
    South of the Clouds, 71–88,
    73, *81, 86, 90*
  expedition stumbles upon a
    lamasery, 60–62
  hardships of the journey,
    53–56, 62, 72–74, 76, 87–88,
    102
  Kermit, Ted, and Suydam
    bathe in the Yalong River, 81
  Kermit shoots a serow, 56–57,
    58, 59–60
  lamas save the expedition, 93
  Roosevelt brothers buy holy
    books but must return them,
    82–83
  Roosevelt brothers hunt big
    game, 83–84
  Tibetan guides for the
    expedition and, 51–52, 53, 57,
    72, 76–78
Hutton, John Henry "J. H.," 43

J

Jade Dragon Snow Mountain
    (Mount Satseto), *48*, 49
  Sackville-West poem, 49–50
Japanese brown bear, *Ursus arctos
    lasiotus*, 96–97
Jenner, Edward, 147

K

Kanai, Shan state, 8, 12–17
Kelley, William Vallandigham,
    xvii
Kingdom of the Golden Monkey,
    118–31
  dense bamboo jungle of,
    124–25
  dogs used in panda hunt,
    126–27
  hunters create a Taoist altar,
    129
  panda scat found, 125, 127
  Roosevelt brothers hunt
    the golden monkey,
    Rhinopithecus roxellana,
    118–24

INDEX

Roosevelt brothers hunt the panda without success, 124–31
Roosevelt brothers reunite with Suydam, 134
Kodiak bear, *Ursus arctos middendorffi*, 97

L

Last Panda, The (Schaller), 206
Lianxian, Han, 205
Lolo land (Yi people)
  death of Hsuen's birds, 141–43
  expedition in the town of Kooing Ma, 153–56, 159
  expedition's encounter with bandits, 142
  expedition's journey to, 141–49
  poppies and mulberry trees of, 144
  Roosevelt brothers' appearance after months of travel, 154, 155
  Roosevelt brothers encounter smallpox victims, 147–48
  Roosevelt brothers' final hunt for the panda, 156–64
  Roosevelt brothers hear about a panda sighting, 148–49, 154–55
  Roosevelt brothers kill the panda, 165–66, *166*
  Roosevelt brothers skin the panda and crate the skin, 168
  Roosevelts warned to stay away from, 101, 139–40
  Yi hunters agree to track the panda, 155–62
  Yi hunters reluctant to participate in killing the panda, 162
  See also Yi people
Lost Horizon (Hilton), 59

M

Matthiessen, Peter, 67, 90
Mekong River, 38
  the Crim and, 37, 38
  dolphins of, 39–41
  expedition in 2021 and, 205
  Kermit and, 38, 39
  water of, 39, 40
Milne-Edwards, Henri, 119
Milner, Joseph, xv, 72
Minya Konka mountain, China, 85, 191–92
Morgan, J. P., xviii
mulberry trees, *Morus alba*, 144
Muli, Kingdom of, 58–59, 62–71
  appearance of the city, 62, *63*
  expedition's host, Chang Ta Li, 65–66, 67, 70, 71, 72
  Herbert arrives, far behind the expedition, 145–47
  House of the Prince, 63–66, *64*, 70, 146
  Kermit's feelings upon entering, 62

253

INDEX

Muli, Kingdom of *(cont.)*
  king Xiang Cicheng Zhaba, 66, 146
  lamastery and lamas, 68, 69, 146
  people called Hsifan, 59
  photograph of Chang Ta Li, his son, and the Roosevelts, *71*
  poverty of the region, 69–70
  remote and inaccessible, 67
  Roosevelt brothers bring gifts, 59, 63, 65, 68
  Roosevelt brothers' status as meaningless, 67
  Ted given a painting of the Wheel of Life, 63, 70, 195
  Tibetan New Year in, 68–69
Muping, 102–3, 108, 114–17
  Chinese killing or evicting the lamas of, 115
  Chinese magistrate as ruler, 114, 116
  description of, 114
  as feudal kingdom, 114
  plight of the Tibetan widow, 115–16
  Roosevelt brothers and Suydam arrive, 114
  Roosevelt brothers and Suydam dine with a surprising dish served, 116–17
  trail to, 111

N

Naga people, Northeast India, 43–45, *44*
Nashi people, 133, *134*
  porter carrying a Chinese man, *107*
New York Zoological Society, 199–200
  Kermit's position with, 196, 197
  mission to advance wildlife conservation, 196

P

palanquins (porter-carried travel coach), 142
Personal narrative of a pilgrimage to el Medinah and Meccah (Burton), 84
Pliny the Elder, 40
polar bear, *Ursus maritimus*, xiii, 98–99
Polo, Marco, xviii, 10
Pride and Prejudice (Austen), 77, 139
Ptolemy II, King of Egypt, xiii

R

red panda, *Ailurus fulgens*, 35
rhododendrons, 71, 76, 77, 104, 155
Rock, Joseph, 59, 71, 105, 138
  describes the king of Muli, 146
Rockefeller, John D., xviii

INDEX

Roosevelt, Archie, 11
Roosevelt, Eleanor (Ted's wife), 11, 101, 195
   Ted's illness in Saigon, and 177–78
Roosevelt, Franklin Delano (FDR), 182
Roosevelt, Kermit, *xii*
   addiction to paraldehyde, 201
   advocacy for a duck hunting ban, 198
   advocacy for endangered species, 189, 196, 198
   advocacy for a panda hunting ban, 198, 199
   alcoholism of, 182, 201, 202
   AMNH expeditions with his father TR, xv–xvi
   attitude of privilege and entitlement, 29
   Audubon Society president, 197–98
   Central Asia expedition of 1925, 10–11
   character and personality, 3, 123
   close relationship with Ted, 31, 173–74
   correspondence with the Chinese, to protect the panda, 198, 199
   dedication to wildlife conservation, 196, 198, 201
   desire to obtain a giant panda, xii
   deterioration of his health, 201–2
   emotional fragility of, 179
   estrangement from Ted, 182
   exploration and, xi
   FDR and, 182
   finances of, xviii, 182
   identity as the president's son, 67
   journal of, 48
   killing of an elephant calf, xvi
   larger-than-life persona, 11
   life-changing impact of the panda expedition, 41, 67–68, 89–90, 116, 167–69, 195, 203
   love of books, 77
   New York Zoological Society position, 196, 197
   nightmares of, 156
   payment for the expedition, 191
   personal problems, 11, 68
   post-expedition misery, 195
   protection of the Sitka spruce, 198
   regrets killing the panda, 179, 195–96, 203
   reunion with Suydam aboard the Astor yacht, 196–97
   saving the Galápagos tortoise, with Suydam, 189, 196–97
   speech about the expedition, 179
   suicide at Fort Richardson, Alaska, 202–3
   Suydam's friendship, 10, 22
   Trailing the Giant Panda, 180
   TR's advice on hunting, 20, 21

INDEX

Roosevelt, Kermit *(cont.)*
   TR's Brazilian exploration of 1913 and, 22, 30, 156
   TR's relationship with, 68, 150–51, 164
   World War II and, 201–2
   writing of Jane Austen and, 77, 203
   See also Roosevelt brothers' giant panda expedition (1928–1929); *specific places on the expedition's itinerary*
Roosevelt, Quentin, 182
Roosevelt, Theodore, Jr. "Ted," vii, *xii*
   acquisition of material possessions, 65, 70, 80, 82, 92, 138–39, 174
   attitude of privilege and entitlement, 29
   bigotry of, 78
   book editing and career at Doubleday, Doran, 194–95
   Central Asia expedition (1925), 5
   close relationship with Kermit, 31, 173–74
   Colonial Policies of the United States, 195
   death in France (1944), 204
   desire to obtain a giant panda, xii
   estrangement from Kermit, 182
   exploration and, xi
   failed political career, 11, 68, 182
   finances of, xviii
   house built by and contentment of, 195
   identity as the president's son, 67
   larger-than-life persona, 11
   life-changing impact of the panda expedition, 41, 67–68, 89–90, 116, 167–69, 195, 203
   life lived with little regard for the consequences, 83
   love of books, 77, 194
   payment for the expedition, 191
   post-expedition distress, 182
   reaction to Harkness's panda cub, 182, 183
   statement upon hearing of Kermit's suicide, 203
   stock market crash and financial peril, 178–79
   Taps: Selected Poems of the Great War, 194
   Teapot Dome scandal, 11
   Trailing the Giant Panda, 180
   TR's relationship with, 150–51, 164, 195
   Wheel of Life painting and, 64–65, 70, 195
   wife, Eleanor and, 11, 101, 177–78, 195
   World War I and, 11, 178
   World War II and, 203–4
   See also Roosevelt brothers' giant panda expedition

INDEX

(1928–1929); *specific places on the expedition's itinerary*

Roosevelt, Theodore (TR), xi
  advises Kermit on hunting, 20, 21
  AMNH expeditions, xv–xvi
  Brazilian expedition of 1913, toll of, 22
  on Chinese coolies, 23
  creation of the Teddy bear, 98
  death of, 22
  exuberance and rugged persona, 11
  game of bear hunt with Ted and Kermit, 150, 164
  killing of endangered species, xvi
  love of books, 77
  need to justify his expeditions, 21
  relationship with his sons, 150–51
  Summer White House, 193
  wealth and financial losses, xviii
  on wildlife preservation by sportsmen, 179
  as women's rights advocate, 57

Roosevelt brothers' Central Asia expedition (1925), xviii–xix, 5, 10
  legendary bighorn (*Ovis poli*) found, xviii

Roosevelt brothers' giant panda expedition (1928–1929)
  announcement of, 3–4
  appearance of the brothers after months of travel, 155, 174, 178
  aura surrounding the Roosevelts and, 5–6, 11
  books carried by the Roosevelt brothers, 77, 84, 139
  "daily annoyances" and "trivialities," 57
  dangers and risks, 6
  dog "Bob" joins the expedition, 51–52, 75
  equal pay for male and female guides, 57
  expedition divides, Roosevelts go to Lolo land, Suydam and Jack head to Yunnanfu, 143
  expedition maps unhelpful, 149
  final phase planned for Southeast Asia, 172–74, 176
  funding for, xvii, xviii, 56, 101, 107
  guide named Luzon, 78
  guide swap in Likiang, 50–51
  hardships of the journey, 50, 53–56, 60, 62, 72–74, 76, 87–88, 125–26, 130, 141–42, 156–57, 162–63
  Hendee to join final phase of the journey, but tragedy intervenes, 172, 176–77
  Herbert causes frustrations, 25, 26–27, 39
  Herbert disappears, and the Roosevelt brothers meet his rescuer, 2, 4–9, 12–17

INDEX

Roosevelt brothers' giant panda expedition (1928–1929) *(cont.)*
    Herbert left behind to rendezvous at Forge of Arrows, 46–47, 145
    Herbert's expedition account: "Sketches of the Tatsienlu Peaks," 189
    Herbert sorts the specimens brought home, 190
    hierarchy of the expedition, 122
    Hsuen, favorite Tibetan guide, 141, 142–44, 147, 148–49, 158, 160, 170, 174
    importance of, to the brothers, 101, 121
    Kermit leaves the expedition, impact on Ted, 173–74
    Kermit's doubts and gloomy forebodings, xix, 21, 30–31, 52
    Kermit shoots a serow, 58, 59–60
    Kermit's illnesses and health, 30–31, 53, 54, 55–56, 62, 87–88, 156–57, 158, 171
    Kermit's jacket burnt and patched, 55–56, 60, 61–62
    lack of communication with the outside world, 171
    members of the expedition, 2, 3, 4, 5
    newspaper reports on the success of the expedition, 179
    number of birds, mammals, and plants collected, 143, 190
    order of caravan during movement, 24
    packs carried by the brothers and porters, 107
    perils that the brothers' position in society could not prevent, 29
    photograph of the caravan, 46
    plant collecting and, 105–6
    rainy season in April as a deadline, 102, 141, 158–59
    Roosevelt brothers' account of: *Trailing the Giant Panda*, 180
    Roosevelt brothers' final hunt for the panda and killing it, 156–66, *166*
    Roosevelt brothers forever changed by the expedition, 41, 67–68, 89–90, 116, 167–69, 174
    Roosevelt brothers' plan for shooting a panda, 124
    Roosevelt brothers' style of travel, 121–22
    setting up camp, 42
    skinning captured birds in Yunnan, 49–50
    specimen collecting, 42–43, 45, 53, 54, 56
    specimens and samples shipped home, 178
    Ted and Suydam's near-fatal bout with malaria and dysentery, 177–78

INDEX

Ted bit by a monkey, 60–61
Ted's infected leg and illness, 171
Ted's wife Eleanor arranges for the end of the expedition, 178
Tibetan guides, 51–52, 53, 72, 77–78, 141–44, 147, 148, 158, 159
unintended consequences, 179–80, 183
warnings to avoid Lolo land, 101, 139–40
women as guides and porters, 22–24, 27–28, 50–51, 188
the "worst night" of their lives, 88, 89–90
Roosevelt brothers' giant panda expedition (1928–1929): itinerary, x
Complete Heaven, then Muping, 102–17
Kingdom of the Golden Monkey, 118–31
Forge of Arrows (Tatsienlu), 89–101
The Happy Valley, 1–18
Himalayan ascent, 38, 42, 47–58
Himalayan trail, South of the Clouds, 71–88
Kingdom of Muli, 58–59, 62–71
Lolo land and the Yi, 140, 141–45, 147–49
pool of the Crim and the dolphins of the Mekong, 37–41
Valley of Death, 19–29
Yachow and the Temple of Hell, 132–39, 141
Yunnanfu, 171–74
See also specific stops on the itinerary
Roosevelt's barking deer, *Muntiacus rooseveltorum*, 204–5
Royal Geographical Society, London, publisher, Herbert's "Sketches of the Tatsienlu Peaks," 189–90

S

Sackville-West, Vita, 49–50
Saigon, Vietnam
Ted and Suydam learn of Hendee's death, 176
Ted's and Suydam's near-fatal bout with malaria and dysentery, 177–78
train to, 174
sambar, *Rusa unicolor*, 56, 84–85
sarus crane (Antigone crane), 17–18, 204
Saw Bwa Fang Tao (Philip Tao), 7–8, 12
family of, 13
fate of his mother, 15–16
his brother's opium addiction 14–15
Kermit names his son, 16–17

INDEX

Saw Bwa Fang Tao (Philip Tao) *(cont.)*
   Roosevelt brothers' expedition and, 12–17
   Roosevelt brothers' gift from, 17
Saw Bwa Fang Yu-chi, 7, 14
Schaller, George, 205
The Last Panda, 206
sea otters, 199
serow (wild goat), 56–57
   Kermit shoots for the Chicago Field Museum, 58
silkworms, *Bombyx mori*, 144–45
   legend of the empress Leizu, 144
smallpox, 147–48
spruce trees of the Himalayas, 79
Stevens, Herbert, 2
   accompanies Suydam to India (1930), 190
   as advocate for habitat protection, 26
   background, 5
   detailed maps of the Himalayas drawn by, 189
   expeditions after the Roosevelt brothers' panda expedition, 190
   final expedition, Papua New Guinea, 190
   heroic entomologists and, 26
   interest in insects and ecosystems, 26–27
   love for land of the lamas, 190
   as magnet for bad luck, 26–27
   retirement and death, 190
   Royal Geographical Society in London and, 189–90
   scientific papers of, 25–26
   "Sketches of the Tatsienlu Peaks," 189
   Through Deep Defiles to Tibetan Uplands: The Travels of a Naturalist from the Irrawaddy to the Yangtse, 190
   vast collection of, 189, 190
   Vietnam expedition (1924), 26–27
   wife, Amy, 26–27
Stevens, Herbert: Roosevelt brothers' panda expedition
   disappearance on the first day, 2
   distractions and delays caused by, 25, 145, 171
   encounters Antigone cranes, 17–18, 204
   fails to rendezvous in Forge of Arrows (Tatsienlu), 103–4
   in Kanai, 8, 12, 13, 15
   location outside of Muli, 104
   Mekong River and the Irrawaddy dolphin, 40
   motive for joining the panda expedition, 5
   in Muli and interaction with the king, 145, 146–47
   new species of butterfly and, 25
   parts ways with the main expedition, 46–47
   photographs by, *42, 46, 48*

INDEX

rescue by Saw Bwa Fang Tao, 7–8
risks of solo traveling, 46–47
sketching and recording birds and animals, 40, 45, 145–46
wandering the high peaks of the Himalayas, 171, 189–90
wandering to the west of the Muping, 118

T

takin (species of goat), 129, 159
 habitat protection for, 206
Taps: Selected Poems of the Great War (Ted Roosevelt), 194
tea
 ancient trees near Yachow, 137
 imported to Tibet, 106
 origins of the world's tea leaves, 137
 porters, *106*, 106, 107, 108, 142
 porters' opium addiction, 108
Temple of Hell, 135–36
Through Deep Defiles to Tibetan Uplands: The Travels of a Naturalist from the Irrawaddy to the Yangtse (Stevens), 190
Tibet
 barley grown in, 80
 Buddhism on redemption, 186
 Buddhist Wheel of Life, 63–65
 Chinese executions in, 133
 customs and food of, 52

Dalai Lama, 52, 187, 188
 expulsion of the Chinese, 52
 forced treaty with the British, 186–87
 guides for the brothers' panda expedition, 51–52, 53, 72, 77–78, 141–44, 147, 148, 158, 159
 Herbert's love for, 190
 Jack's description of Tibetans, 91
 lamas and lamaseries of, 60–62
 Lhasa apsos breed of dog, 188
 New Year or Losar, 68–69
 pink salt of, 37
 Roosevelt expedition granted a rare permit to visit, 187
 Suydam's return in search of the sacred, 187–88
 tea porters and, 106, *106*
 tea with yak butter, 53, 61
 terma or treasure texts, 83
 tsampa as diet staple, 80
 tsampa bowl, 79–80
 typical greeting, 63, 82
 women in, 188
Tibetan Book of the Dead, The (Bardo Thodol Chenmo), 83, 89
Tibetan Plateau, 38, 62, 105, 120, 160, 164, 167
 golden snub-nosed monkey and, 119–20
Trailing the Giant Panda (Ted and Kermit Roosevelt), 180

INDEX

V

Valley of Death, 19–37
- difficulties of the journey, 30
- expedition's encounter with bandits, 27–28
- Herbert and a new butterfly species, 25
- Jack finds a doctor for Kermit, 31–33
- Kermit and Ted make inquiries about a panda, 33–36
- Kermit shoots at a river otter, 19–21
- Kermit's illness, 30–31
- Kermit's trepidation, 21, 22–23
- pelts of the red panda found, 34–35
- Roosevelt brothers are unappreciative of the perils, 28–29
- Suydam's "sense of oppression," 31
- Ted's panic about Kermit's illness, 31
- warning about evil spirits, 21, 27

Vientiane, Laos, 176
- Hendee's death and headstone, 176–77

W

Waterman, Henry S., 178
Wilson, Ernest, 109
World War II
- D-Day, 203–4
- Jack as a hero in, 194
- Kermit's role in, 201–3
- Suydam and, 188–89
- Ted's role in, 203–4

Y

Yachow (Ya'an China), 103, 132–39, 141
- Baptist mission and Dr. Crook, 137–38, 139–40
- description of, 137–38
- executions by the Kuomintang, witnessed by Jack, 132–34
- expedition mule caravan headed by Jack sets off to, 103
- fighting with the Kuomintang in the countryside, 136–37
- Kermit watches opium distillers, 139
- region of ancient tea trees, 137
- Roosevelt brothers and Suydam set out for, encounter bees, 135, 136–37
- Roosevelt expedition prepares for the most dangerous leg of the journey, 139
- Roosevelt expedition stays in, 137–40
- Ted shops for treasures, 138–39
- telegrams awaiting the expedition, 138
- warning about Lolo land, 139–40

INDEX

Yi people (Lolo), 140
  caste system, 140
  clothing of, 157
  disease and, 158
  group of Yi men, *151*
  as insular society, 140
  location of, 140
  matriarch, Vooka, 169–70
  missionaries rejected by, 154
  panda considered a demi-god and not hunted, 152, 159
  payment to the Yi hunters, 156
  photograph, bamboo house in Yi village, *153*
  response to the Roosevelts' killing the panda, 169
  Roosevelt brothers' first encounter and gift of a hat, 151–52
  Roosevelt brothers offer gifts in Kooing Ma, 154
  Roosevelts shower the village with presents, 170
  slavery and, 140
  why they agreed to send hunters with the Roosevelts to track the panda, 155–56
  See also Lolo land
Young, Adelaide Su-Lin Chen, 193
Young, Jolly, 194
Young, Quentin, 181
Young, Tai Jack "Jack," 2, 3
  accepted as a student at the University of Chicago, 191
  ambitions of, 32
  AMNH funds his China and Tibet expedition of 1937, 193
  appearance, 3
  aura of the Roosevelts and, 11
  changes his name in tribute to the Roosevelts, 191
  character and personality, 100
  Chiang Kai-shek and his family's losses, 16
  Chinese mother of, 32, 37
  death of, 194
  expedition of 1932 to climb Minya Konka, 191–92
  Explorers Club and, 191
  how and why he joined the Roosevelts' expedition, 4, 5
  hunting in Asia for Chinese museums, killing a panda, 192
  influence of the expedition on his life, 191
  as interpreter and guide, 3, 4
  intervention of war as a theme in his life, 192
  Japanese bombing of Shanghai in 1932 and, 192
  knowledge of languages, 32, 91
  marries Chinese American woman, Adelaide Su-Lin Chen, 193
  as mixed race, prejudices felt by, 32, 192–93
  payment for the expedition, 191
  post-expedition life, 100, 187, 191–94

Young, Tai Jack "Jack," *(cont.)*
  relationship with Ted, 192
  Roosevelt brothers and
    Suydam finance his China
    expedition (1934), 193
  as the Roosevelt Summer
    White House, 193
  World War II and being
    drafted into the Chinese
    army, 193–94
  World War II and his heroic
    actions in the US Army, 194
Young, Tai Jack "Jack":
    Roosevelt brothers' panda
    expedition
  description of people and
    places encountered, 91
  encounter with bandits and
    help from the Chinese army,
    111–14
  expedition divides, he and
    Suydam go on alone to
    Yunnanfu, 143
  expedition mule caravan
    arrives in Yachow, Jack
    witnesses executions, 132–34
  expedition mule caravan to
    Yachow left in his charge,
    103, 111–14, 118
  hunting in the Himalayas, 82,
    84, 99–100
  meeting with Saw Bwa Fang
    Tao, 7–9, 16
  misgivings heading into the
    Valley of Death, 21
  in Muli, 68–69, 72
  searching for Herbert, 4–7
  seeks help for ailing Kermit,
    31–33
  specimen collecting, 53
  women porters' conversation
    with, 24
Younghusband, Sir Francis,
  186
Yunnanfu, city of, 171–74
  breakup of the group in, 174
  lake of, 171
  photograph of Ted, Suydam,
    Kermit, and Culver B.
    Chamberlain, *172*
  Roosevelt brothers arrange
    for panda skin to be shipped
    home, 171
  Roosevelt brothers reunite
    with Suydam and Jack, 171
  Ted gives Hsuen his rifle, 174
  urgent cables for Kermit, and
    his departure, 173

ABOUT THE AUTHOR

Nathalia Holt, PhD, is the *New York Times* bestselling author of *Rise of the Rocket Girls*, *The Queens of Animation*, and *Cured: How the Berlin Patients Defeated HIV and Forever Changed Medical Science*. She has written for numerous publications including the *New York Times*, the *Wall Street Journal*, the *Los Angeles Times*, *The Atlantic*, *Slate*, *Popular Science*, and *Time*, as well as for PBS. She is a former fellow at the Ragon Institute of Massachusetts General Hospital, MIT, and Harvard University. She lives with her husband and their two daughters in Pacific Grove, California.